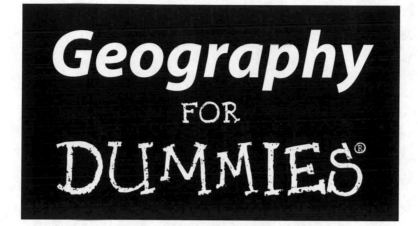

Geography FOR DUMMIES®

by Charles Heatwole

Foreword by Ruth I. Shirey

Hungry Minds™

Best-Selling Books • Digital Downloads • e-Books • Answer Networks • e-Newsletters • Branded Web Sites • e-Learning

New York, NY ◆ Cleveland, OH ◆ Indianapolis, IN

Geography For Dummies®

Published by:
Hungry Minds, Inc.
909 Third Avenue
New York, NY 10022
www.hungryminds.com
www.dummies.com

Library of Congress Control Number: 2002100155

ISBN: 0-7645-1622-1

10 9 8 7 6 5 4 3 2 1

1O/SY/QS/QS/IN

Distributed in the United States by Hungry Minds, Inc.

Distributed by CDG Books Canada Inc. for Canada; by Transworld Publishers Limited in the United Kingdom; by IDG Norge Books for Norway; by IDG Sweden Books for Sweden; by IDG Books Australia Publishing Corporation Pty. Ltd. for Australia and New Zealand; by TransQuest Publishers Pte Ltd. for Singapore, Malaysia, Thailand, Indonesia, and Hong Kong; by Gotop Information Inc. for Taiwan; by ICG Muse, Inc. for Japan; by Intersoft for South Africa; by Eyrolles for France; by International Thomson Publishing for Germany, Austria and Switzerland; by Distribuidora Cuspide for Argentina; by LR International for Brazil; by Galileo Libros for Chile; by Ediciones ZETA S.C.R. Ltda. for Peru; by WS Computer Publishing Corporation, Inc., for the Philippines; by Contemporanea de Ediciones for Venezuela; by Express Computer Distributors for the Caribbean and West Indies; by Micronesia Media Distributor, Inc. for Micronesia; by Chips Computadoras S.A. de C.V. for Mexico; by Editorial Norma de Panama S.A. for Panama; by American Bookshops for Finland.

For general information on Hungry Minds' products and services please contact our Customer Care department; within the U.S. at 800-762-2974, outside the U.S. at 317-572-3993 or fax 317-572-4002.

For sales inquiries and resellers information, including discounts, premium and bulk quantity sales and foreign language translations please contact our Customer Care department at 800-434-3422, fax 317-572-4002 or write to Hungry Minds, Inc., Attn: Customer Care department, 10475 Crosspoint Boulevard, Indianapolis, IN 46256.

For information on licensing foreign or domestic rights, please contact our Sub-Rights Customer Care department at 212-884-5000.

For information on using Hungry Minds' products and services in the classroom or for ordering examination copies, please contact our Educational Sales department at 800-434-2086 or fax 317-572-4005.

Please contact our Public Relations department at 212-884-5163 for press review copies or 212-884-5000 for author interviews and other publicity information or fax 212-884-5400.

For authorization to photocopy items for corporate, personal, or educational use, please contact Copyright Clearance Center, 222 Rosewood Drive, Danvers, MA 01923, or fax 978-750-4470.

Hungry Minds™ is a trademark of Hungry Minds, Inc.

About the Author

Charles Heatwole is Professor of Geography and Chairperson of the Department of Geography at Hunter College of CUNY. He holds a B.A. in Social Studies Education from Florida Atlantic University, and M.A. and Ph.D. degrees in Geography from Michigan State University. In between those schools he served as a Peace Corps Volunteer in West Africa. He attributes his affinity for geography to a childhood passion for stamp collecting, frequent family relocations pursuant to his father's work as a defense contractor's field representative, and a superb high school geography teacher.

Charlie's research has focused on cultural geography, and especially the geography of religion. He has also been involved in geographic education, serving five years as co-coordinator of the New York Geographic Alliance — a teacher training network affiliated with the National Geographic Society's Geography Education Program. In that, and subsequent capacities, he has helped organize and conduct numerous workshops and institutes devoted to the teaching of geography.

Charles lives in Manhattan with his wife, Debbie, and daughter, Mary. He enjoys jogging and, of course, travel. His favorite professional activity is teaching Geography 101.

Dedication

In grateful memory of the 343 members of the New York City Fire Department who perished in the events of September 11, 2001 and two very dear people from the National Geographic Society, Mr. Joe Ferguson and Ms. Ann Judge, who were aboard the airliner that struck the Pentagon.

Author's Acknowledgments

This is the place where authors wax humble about how they couldn't have done *blah blah blah* without the help of Tom, Dick, and Harry. Well, there's a reason for that: they couldn't have.

Thanks must go first of all to Carolyn Krupp, my agent at IMG Literary. It was she who telephoned out of the blue to tell me about this project and why I was the person to do it. To this day I have no idea how she got my name or number. In any event, "Thanks, Carolyn."

At Hungry Minds, Inc., thanks also go to Linda Brandon, my editor, and Roxane Cerda, the acquisitions editor for this book. Both ladies were extremely helpful, encouraging, and patient when I needed it most. Through the magic of modern electronics, I was able to complete this book without ever meeting Linda and Roxanne face-to-face. Indeed, to this day I don't know what either of them looks like. Ladies, I hope we meet someday. At the very least, I owe you lunch.

I thank everybody at Hunter College who indulged me in ways great and small while I was writing this. In my case at least, writing a book and being department chairperson proved incompatible. Something had to give; and more often than not it was my professional duties. Special thanks, therefore, are due to Anthony Grande, Assistant to the Chair, for doing so many little things that added up to a lot of time for me to work on this. Thanks also to my departmental colleagues Allan Frei, Ines Miyares, and Randye Rutberg for supplying several essential tidbits of information.

It has been my privilege to be acquainted with a number of outstanding K-12 teachers. Among other things, they taught me how much easier it is to borrow an idea than to invent one. I suspect this manuscript contains several quips and ideas which, though they popped up in my mind, probably originated in one of theirs. Proper attribution escapes me, so I'll just say, "Thanks to you all."

I want to thank all of the students I have had in my classes at Hunter College over the years. So much of this book is an outgrowth of classroom experience — things you say and do in class as a result of years of trial and error. If this book communicates effectively in whole or part, then much credit is due to the thousands of student guinea pigs who sat through my lectures.

Finally and most of all, thanks to my wife and daughter, Debbie and Mary, for being so loving, supportive, and basically putting up with this. They spent a lot of time together while daddy hunkered down. I was away too long.

Publisher's Acknowledgments

We're proud of this book; please send us your comments through our Online Registration Form located at www.hungryminds.com

Some of the people who helped bring this book to market include the following:

Acquisitions, Editorial, and Media Development

Project Editor: Linda Brandon

Acquisitions Editor: Roxane Cerda

Copy Editor: Robert Annis

Technical Editor: Prof. Thomas Tharp

Senior Permissions Editor: Carmen Krikorian

Editorial Manager: Christine Beck

Editorial Assistant: Melissa Bennett

Cover Photos: ©Corbis

Production

Project Coordinator: Erin Smith

Layout and Graphics: Jackie Nicholas, Jacque Schneider, Julie Trippetti, Jeremey Unger

Proofreaders: Laura Albert, TECHBOOKS Production Services

Indexer: TECHBOOKS Production Services

Hungry Minds Consumer Reference Group

Business: Kathleen Nebenhaus, Vice President and Publisher; Kevin Thornton, Acquisitions Manager

Cooking/Gardening: Jennifer Feldman, Associate Vice President and Publisher; Anne Ficklen, Executive Editor; Kristi Hart, Managing Editor

Education/Reference: Diane Graves Steele, Vice President and Publisher

Lifestyles: Kathleen Nebenhaus, Vice President and Publisher; Tracy Boggier, Managing Editor

Pets: Kathleen Nebenhaus, Vice President and Publisher; Tracy Boggier, Managing Editor

Travel: Michael Spring, Vice President and Publisher; Brice Gosnell, Publishing Director; Suzanne Jannetta, Editorial Director

Hungry Minds Consumer Editorial Services: Kathleen Nebenhaus, Vice President and Publisher; Kristin A. Cocks, Editorial Director; Cindy Kitchel, Editorial Director

Hungry Minds Consumer Production: Debbie Stailey, Production Director

Contents at a Glance

Cartoons at a Glance

By Rich Tennant

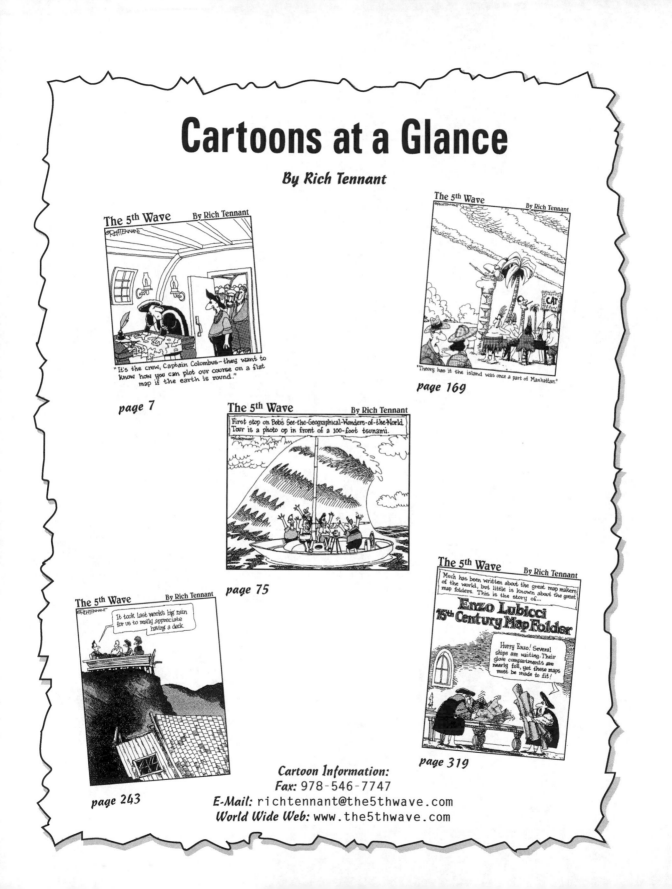

The 5th Wave — By Rich Tennant

"It's the crew, Captain Columbus—they want to know how you can plot our course on a flat map if the earth is round."

page 7

The 5th Wave — By Rich Tennant

"Theory has it the island was once a part of Manhattan."

page 169

The 5th Wave — By Rich Tennant

First stop on Bob's See-the-Geographical-Wonders-of-the-World Tour is a photo op in front of a 100-foot tsunami.

page 75

The 5th Wave — By Rich Tennant

It took last week's big rain for us to really appreciate having a deck.

page 243

The 5th Wave — By Rich Tennant

Much has been written about the great map makers of the world, but little is known about the great map folders. This is the story of...

Enzo Lubicci
15th Century Map Folder

Hurry Enzo! Several ships are waiting. Their glove compartments are nearly full, yet these maps must be made to fit!

page 319

Cartoon Information:
Fax: 978-546-7747
E-Mail: richtennant@the5thwave.com
World Wide Web: www.the5thwave.com

Table of Contents

Part IV: Putting the Planet to Use 243

Chapter 15: Getting Down to Business 245

Chapter 16: An Appetite for Resources 263

Foreword

∙∙∙

Geography is for life in every sense of that expression: lifelong, life-sustaining, and life-enhancing. Geography is a field of study that enables us to find answers to questions about the world around us — about where things are and how and why they got there. We can ask questions about things that seem very familiar and are often taken for granted.

Geography is the science of space and place on Earth's surface. Its subject matter is the physical and human phenomena that make up the world's environments and places. Geographers describe the changing patterns of places in words, maps, and geographics, explain how these patterns come to be, and unravel their meaning. Geography's continuing quest is to understand the physical and cultural features of places and their natural settings on the surface of Earth.

— Geography Education Standards Project. *Geography for Life: National Geography Standards 1994*. Washington, DC: National Geographic Research & Exploration.

*B*ack in the 1980's professional geographers began to hear something about our field that we had rarely heard before. We began to hear some of our colleagues and acquaintances in other fields — educators, accountants, salespersons, doctors, systems analysts, draftspersons, tax collectors — admit to having liked geography when they were in school. A psychologist at my university told me about the wonderful cultural geography course he had taken as an undergraduate at a major east coast university. Any number of people came out of the closet and admitted that they had always liked to study maps. An anthropologist told me how as a child he had loved his United States puzzle map. Wow! This was a different experience than having people look puzzled, or worse, ask upon meeting you, "Haven't all the places been discovered, located, and named already?"

What was happening, of course, was that the United States was gradually becoming aware that it's citizens appalling lack of geographic knowledge of our own country and other places on Earth was a threat to its future economic, environmental, and political well-being. In a world shrunk by communications and transportation technologies, we must know where places are located, but even more we have to know about those places and the physical and human processes that shape them. At no time has the seriousness of the situation been more evident than on September 11, 2001. Horror and grief at the loss of innocent life binds us together as a people within our borders and beyond. We are reminded of our interdependence at local, regional, and global scales. As we are challenged to understand what happened, the question of "Where?" is at the forefront, again, at local, regional, and global scales.

Geography For Dummies provides the opportunity to gather information from an experienced teacher who is dedicated to solving the geographic literacy problems, which have been part of human culture for just too long. The opportunity is provided in an engaging way that doesn't insult your intelligence. *Geography For Dummies* is an introduction to contemporary analytical geography for the lay person. The emphasis in this book is on geography as a field of inquiry framed by spatial and environmental perspectives about the places, environments, and cultures on Earth's surface. The book reflects traditions and themes about which geographers have achieved a high degree of consensus both before and at the publication of *Geography for Life: National Geography Standards 1994* (quoted previously). *Geography For Dummies* references the six essential elements of geography used as organizing concepts in *Geography for Life*.

Therefore, the content of this book reflects what geographers and geography educators have recommended as the most important geography stuff everyone should know.

— Ruth I. Shirey, PhD in Geography

Executive Director of the National Council For Geographic Education since 1988, Coordinator for the Pennsylvania Geographic Alliance since 1986, and Professor, Department of Geography and Regional Planning at Indiana University of Pennsylvania

Introduction

· ·

*T*he best teacher I ever had taught geography at William R. Boone High School in Orlando, Florida. Her name was Mary Row, and it's a shame you didn't know her as I did — at the very least, it would have saved you the cost of this book.

Mrs. Row had an incredible knack for taking a class of tenth-graders and turning them into ex-dummies, at least with respect to geography. Indeed, some of the students who walked into her classroom on the first day of school simply didn't give a hoot. I know. I was there. But when that lady got done with us, we were upstart experts on geography and loved the subject.

In retrospect, I believe a principal key to Mrs. Row's success was her philosophy of geography. As far as she was concerned, Earth is a very fascinating place. The purpose of geography, as she saw it, is to convey the wonderment of it all and to explain how the world works. Thus, her lessons emphasized the interactions between the various things that characterized Earth's surface and how they related to everyday life. So thanks to Mrs. Row, geography was not only the most interesting subject I studied in high school, but also the most relevant.

Hopefully, this volume will instill in you some measure of the wonderment that came my way those many years ago, and whet your appetite for more.

About This Book

Introductory books on geography generally come in two varieties. This one takes a *topical* approach to the subject. That means the chapters focus on topics of interest to geography, such as maps, climate, population, and culture. I wanted this book to focus on the key concepts of geography and introduce you to a wide-range of geographic information. Basically, I thought those goals could best be achieved by taking a topical approach.

The alternative was to take a regional approach to geography, which is like a world tour. You know what I mean, right? Chapter 4: Europe. Chapter 5: Africa. And so forth. In all candor, I didn't think I could give you a decent world tour in the allotted pages. Besides, books like that for people like you are already on the market, so why reinvent the wheel? More importantly, I wanted *Geography For Dummies* to emphasize geography rather than the world *per se*. That may cause you to say, "Wait a minute! Isn't geography all about the world?" The answer is yes, but in a larger sense, geography is about a whole lot more. Specifically, it's about concepts and processes and connections between things, plus maps and tools and perspectives that combine individual "world facts" and give you big pictures that are so much more meaningful than their myriad components. Parenthetically, there's a curious thing about those geography-as-world-tour books. They all seem to start by telling you geography is so much more than facts about the world, and then spend 350 pages telling you facts about the world.

Foolish Assumptions

I'm going to assume that you are an average person who is curious about the world but who just happens to have a limited background in geography. And I firmly believe "average" means intelligent, so nothing is out of bounds because of the gray stuff between your ears. Instead, in my view, you are completely capable of digesting the real meat and drink (or tofu and carrot juice, if you prefer) of geography. You may be 14, or 44, or 84. It doesn't matter. As far as I am concerned, you're ready for prime-time geography. Please understand I'm not talking wimpy stuff like "What's the capital of Nevada?" No way. I'm talking big league stuff like how you can have a rainforest on one side of a mountain range and a desert on the other; or how to choose a good location for a shopping mall; or how ocean currents help to determine the geography of climates.

I'm also going to assume that, generally speaking, you know your way around the world. Thus, when you see terms like Pacific Ocean, Nile River, Europe, or Japan, some kind of mental map pops up inside your head and allows you to "see" where they are located. On the other hand, when you meet up with terms like Burkina Faso, Skaggerak, or Myanmar, you may need some outside help. For that reason, it will be helpful to have an atlas handy.

Finally, if this were a beer, then I'm assuming you went to your bookstore to pick up some Geography Lite. That is, you want the real thing, but figure you don't need all the calories. One of my goals is to make this book a painless — and indeed a pleasurable — experience. A lite-hearted read, if you will, that also communicates some serious geography and leaves you with a well-rounded exposure to the subject. If that sounds about right, then I invite you to keep reading.

How This Book Is Organized

This book is divided into four major parts that address broad areas of geography, plus a fifth part with ancillary information. Each of these parts consists of chapters that concentrate on an important aspect of that subject area. Following is the full story.

Part 1: Getting Grounded: The Geographic Basics

This section introduces you to the major concepts, modes of thinking, and tools of geography. Sadly, many people think that geography is little more than a category on TV quiz shows. Accordingly, my first task is to set the record straight concerning what geography is and what it is not. Thus, you will encounter examples that highlight the nature of geography and show you how to think like a geographer.

Maps are the most basic tools of geography. If this book didn't talk about them in some detail, then it couldn't claim to be a grown-up primer on geography (which it does). Thus, you encounter an overview of latitude and longitude, the basic principles of map-making, and the fundamentals of map reading. In addition to maps, which are about as old as geography itself, modern geographers use some really neat cutting-edge technology that helps them locate and analyze phenomena on Earth's surface. You'll meet some of that technology in this section.

Part II: Getting Physical: Land, Water, and Air

This section introduces you to Earth's physical characteristics and the processes that underlie them. Geography plays out on an Earthly stage of astonishing variety. Landforms, water bodies, soils, vegetation — they're all here. And above it all is a remarkable atmosphere that gives us air to breathe, rainfall to sustain plant and animal life, and temperature environments that warm us up, chill us out, and do everything else in between.

Therefore, understanding the characteristics and locations of the Earth's natural features is fundamental to a sound geographic education. But landforms and other aspects of the natural world don't "just happen." Everything you see today, everything that existed yesterday, and everything that will characterize Earth in the future are the result of natural processes. Understanding these processes is as fundamental to geography as knowing the landforms they produce; for only by understanding the processes can you really understand the world.

Part III: Peopling the Planet

This section introduces you to the basic content and concepts of human geography. Arguably, people are the most important phenomenon that characterizes Earth's surface, and probably the most complex and diverse as well. Areas of extraordinary population density contrast with regions in which people are few and far between — at least for now. That qualification is appropriate because, thanks to migration and reproductive biology, the distribution of people is forever in a state of flux.

But human geography is not just a numbers game. All humans possess an array of culture traits which, in their depth and breadth, not only differentiate one group of people from the next, but also add substantial variety to the look and feel of the world in which we live. On top of that, people are territorial. They have a propensity to divide and control Earth's surface, creating countries and other political entities that, by creating nationalities and jurisdictions, further characterize and complicate the picture. In occupying the planet, therefore, people have imparted a rich mosaic of attributes to their Earthly home. Acquiring a basic understanding of them is part and parcel of becoming a geographically informed person.

Part IV: Putting the Planet to Use

This section focuses on characteristics and consequences of human use of Earth. As Parts II and III respectively emphasize, natural features (like landforms and climates) characterize our world, and human features do the same. But these sets of characteristics do not exist in isolation from each other. We humans not only occupy the planet — we also put it to use as we construct our homes and settlements, make a living, produce our food, garner resources, and dispose of our refuse.

The Earth, therefore, is a natural entity that we impact and modify. Increasingly, therefore, as geographers describe and explain Earth's character, the story line concerns the role of humans in changing the face of the planet and altering its environmental quality, usually for the worse. Different people impact different regions in different ways. Nevertheless, general principles and concepts have been identified that help us to understand the nature and results of our actions and also hint at strategies to improve our planetary stewardship. You'll be introduced to them in this section.

Part V: The Part of Tens

You want lists? Well, in the grand finale tradition of *For Dummies* books, I give you lists. They concern organizations and agencies that can provide you with very useful information and materials and information about careers in geography. And for a real change, if you feel like forgetting a few things, you can find a list for that, too!

Icons Used in This Book

From time to time you will encounter circular, cartoon-like figures in the left-hand margin of the text. The purpose of these icons is to alert you to the presence of something that is comparatively noteworthy amidst the passing prose. That may be something I regard as particularly important, or something you may wish to take your time to think about, or something you may wish to skip. In any event, here are the icons and their meanings.

This icon identifies a major concept that is "big" in the sense of having widespread applicability or broad explanatory power. It does not necessarily mean "difficult to understand." Indeed, most big ideas turn out to mean something rather simple.

Like many subjects, geography contains some specialized and perhaps arcane vocabulary terms that cause normal, well-adjusted people like you to scratch their heads. I could bypass this geo-jargon altogether, but then you really wouldn't be discovering more about geography, would you? So instead, you and I will meet the jargon head-on and do what is necessary to make sure you understand it. Graticule? Complementarity? Evapotranspiration? They're not going to be a problem.

Occasionally, you will encounter a rule of thumb that clarifies a concept or helps to make sense of something. Likewise, you will sometimes come across a sentence or phrase that captures the essence of a principle or the theme of a chapter or of the entire book. Those kinds of tidbits are especially worth remembering and are identified by this icon.

Geography involves elements of math, science, technology, ecology, modeling, and other technical stuff. Some will show up in this book because they are relevant to a well-rounded geographic education even at this introductory level. I do appreciate, however, that some people may find these a bit too complicated. This icon alerts you to the presence of technical stuff and is meant to suggest three things. First, you have encountered something that is somewhat challenging to comprehend relative to the general contents of this book. Second, it's no reflection on your intelligence if you think this is a bit complicated. Third, you can skip it if you wish. As a side note, this icon has been likened to a nerdy guy with big glasses. I fear it looks a little too much like me.

Some aspects of geography are a little involved, so it's always nice to encounter information that helps you simplify a process or cut through the bafflegab and make things easier to comprehend. Those are the kinds of items this icon pinpoints.

Where to Go from Here

You can take that two ways: where to go when you're done with this page, and where to go when you're done with this book. Regarding the former, I recommend you read this book from start to finish as you would a novel. To some extent, geographic knowledge is cumulative. That is, there are basic concepts and information that provide a foundation for understanding other concepts and information. Accordingly, the parts and chapters of this book follow a certain logical progression. In short, I do believe the content of this book will make more sense to you if you read this volume from start to finish. However, if you wish, you can dive into chapters at random — each chapter is set up to be self-contained.

And where do you go when you're done? As I've already mentioned, the Part of Tens (Part V) contains references to careers in geography and organizations and web sites that can further your interest in and mastery of the subject. So, if in fact this book whets your appetite for more, then there's information at the end to help satisfy the hunger.

Part I
Getting Grounded: The Geographic Basics

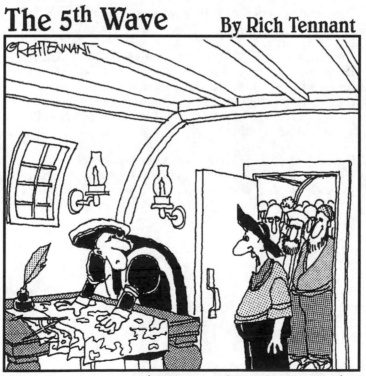

The 5th Wave   By Rich Tennant

"It's the crew, Captain Columbus—they want to know how you can plot our course on a flat map if the earth is round."

In this part . . .

Each and every academic discipline has its own particular and peculiar subject matter. Geography is no exception, but my, how things have changed!

For the longest period, geography was concerned primarily with mapping the world and acquiring facts about places. It has since become a much more analytical pursuit. Thus, the time-honored imperative to know *where* things are located is complemented by an equally strong (if not stronger) desire to know *why* they occur where they do. Also, geography has become an applied discipline that seeks to identify the best location for a hospital, store, factory, or other facility.

In this part, you will learn about the key concepts and methods of contemporary geography as well as the principal tools and techniques of the trade. Among other things you will see how exciting technologies are giving geographers unprecedented perspectives on *where* and *why*.

Chapter 1

Geography: Understanding a World of Difference

You live on a very interesting planet, a world of never-ending variety — mountains and plains, oceans and rivers, deserts and forests. If, as Shakespeare once wrote, "All the world's a stage," then one could hardly imagine a greater range of sets and scenery than exists on planet Earth.

You are an actor on that stage, and you are not alone. The entire cast number more than six billion, and they are as diverse as their Earthly stage. They practice dozens of religions, speak many hundreds of languages, and display thousands of cultures. They live in scattered farmhouses, large cities, and every-size settlement in between. They practice every kind of livelihood imaginable and, in innumerable ways great and small, have interacted with and changed the natural environment forever.

So "interesting planet" and "never-ending variety" turn out to be code for "complex." Truly, this is a complex world in which no two areas are exactly alike. On the one hand, this complexity makes for a very fascinating planet. But on the other hand, the prospect of learning all about this complexity can be overwhelming, or at least sometimes seems to be. Fortunately, one subject seeks to make sense of it all and, usually, does a pretty good job: Geography.

Geography: Making Sense of It All

People are fascinated by the world in which they live. They want to know what it's like and why it is the way it is. Most importantly, they want to understand their place in it. Geography satisfies this curiosity and provides practical knowledge and skills that people find useful in their personal and professional lives. This is nothing new.

From ancient roots . . .

Geography comes from two ancient Greek words: *ge,* meaning "the Earth," and *graphe,* meaning "to describe." So, when the ancient Greeks practiced geography, they described the Earth. Stated less literally, they noted the location of things, recorded the characteristics of areas near and far, and used that information in matters of trade, commerce, communication, and administration.

Disputed paternity

A Greek named Eratosthenes (died about 192 B.C.) is sometimes called the "Father of Geography" since he coined the word "geography." The Greeks themselves called Homer the "Father of Geography" because his epic poem, *Odyssey,* written about a thousand years before Eratosthenes was born, is the oldest account of the fringe of the Greek world. In addition to these gentlemen, at least two other men have been named "Father of Geography," all of which suggests a very interesting paternity suit. But I digress. That the story goes back to the days of the Greeks tells us that geography is a very old subject. People of every age and culture have sought to know and understand their immediate surroundings and the world beyond. They stood at the edges of seas and imagined distant shores. They wondered what lies on the other side of a mountain or beyond the horizon. Ultimately, of course, they acted upon those speculations. They explored. They left old lands and occupied new lands. And as a result, millennia later, explorers like Columbus and Magellan found humans almost everywhere they went.

Links to exploration

Geographers from ancient Greece through the 19th century were largely devoted to exploring the world, gathering information about newfound lands, and indicating their locations as accurately as possible on maps. Sometimes the great explorers and thinkers got it right, and sometimes they did not (see the sidebar called "Measuring the Earth"). But in any event, geography and exploration became intertwined; so, "doing geography" became closely associated with making maps, studying maps, and memorizing the locations of things (see Chapters 3 through 5 for information on locating things and creating and reading maps).

TECHNICAL STUFF

Measuring the Earth

In the third century B.C., the Greek scholar Eratosthenes made a remarkably accurate measurement of Earth's circumference. At Syene (near Aswan, Egypt), the sun illuminated the bottom of a well only one day every year. Eratosthenes inferred correctly this could only happen if the sun were directly overhead the well — that is, 90° above the horizon. By comparing that sun angle with another one measured in Alexandria, Egypt, on the same day the sun was directly overhead at Syene, Eratosthenes deduced that the distance between the two locations was one-fiftieth (1/50th) of Earth's circumference. Thus, if he could measure the distance from Syene to Alexandria and multiply that number times 50, the answer would be the distance around the entire Earth.

There are diverse accounts of the method of measurement. Some say Eratosthenes had his assistants count camel strides (yes, camel strides) that they measured in *stade*, the Greek unit of measurement. In any event, he came up with a distance of 500 miles between Syene and Alexandria. That meant Earth was about [500 x 50 =] 25,000 miles around. ("About" because the relationship between stade and miles is not exactly known.) The actual average circumference is 24,680 miles so Eratosthenes was very close.

About a century-and-a-half later, another Greek named Posidonius calculated Earth's circumference and came up with 18,000 miles. Posidonius' measurement became the generally accepted distance thanks to Strabo, the great Roman chronicler, who simply did not believe that the Earth could be as big as Eratosthenes said it was. About 18 A.D. Strabo wrote his *Geography*, which became the most influential treatise on the subject for more than a millennium. *Geography* credited the calculations of Posidonius and rejected those of Eratosthenes. And that leads to an interesting bit of speculation. Columbus was familiar with *Geography*, so he was aware of the official calculation of Earth's circumference — 18,000 miles. Had he known the true circumference was 25,000 miles, like Eratosthenes said, Columbus would have known that China was thousands of miles farther to the west than Strabo suggested. And if he had known the true distance to China, would Columbus ever have set sail?

... *To modern discipline*

During the past century, and especially during the past several decades, geography has blossomed and diversified. Old approaches that focused on location and description have been complemented by new approaches that emphasize analysis, explanation, and significance. On top of that, satellites, computers, and other technologies now allow geographers to record and analyze information about the Earth to an extent and degree of sophistication that were unimaginable just a few years ago.

As a result, modern geographers are into all kinds of stuff. Some specialize in patterns of climate and climate change. Others investigate the distribution of diseases, or the location of health care facilities. Still others specialize in urban and regional planning, or resource conservation, or issues of social justice, or patterns of crime, or optimal locations for businesses. . . . — the list goes on and on. Certainly, the ancient *ge* and *graphe* still apply, but geography is much more than it used to be.

Exposing Misconceptions: More Than Maps and Trivia

Geography is a widely misunderstood subject. Many people believe it's only about making maps, studying maps, and memorizing locations. One reason is that polls and pundits occasionally decry the "geographic ignorance" of Americans, which usually means the average person doesn't know where important things are located. Presumably, therefore, if you memorize the world map, then you "know geography." Another reason is that on many TV quiz shows, contestants are occasionally asked "geography questions." Almost always, the answer is a fact that can be understood by studying a map and/or memorizing the locations of things or events.

Knowledge of the location of things is important and useful. Everything happens somewhere; and if you know the where, then the event has meaning that it otherwise would not. So map memorization is cool, but you need to keep it in perspective. Memorizing locations is to geography what memorizing dates is to history, or what memorizing the multiplication table is to mathematics. Namely, it's a foundation — a base — upon which you can build and develop deeper understandings. The bottom line is: There is more to geographic awareness than *whereness*. And the goal of this book is to uncover how to get beyond whereness when finding out about geography.

In what country is New Mexico?

New Mexico bills itself as "The Land of Enchantment." That slogan is written on their license plates. Or rather, it used to be. Now the license plates say, "New Mexico, USA." Geographic ignorance is the reason for the change. Sadly, many Americans do not know that New Mexico is one of the 50 States. They figure the name refers to that country south of Texas. Think I'm kidding? Some New Mexicans can tell you geographic horror stories. Take the high school student who applied for admission to a Midwestern university, only to be told that his application had to be re-routed through the foreign student office because, well, New Mexico is a foreign country. Naturally, tourism and business development can't help but suffer if Americans don't know that New Mexico is part of their country. So the New Mexicans have seen to it: good-bye "Land of Enchantment" and hello "New Mexico, USA."

Taking a Look at the New Geography

Geographers still make maps and study them, and certainly, geography still consists of subject matter that cries out to be memorized. But the "old geography" of map memorization and descriptive studies has been complemented by a "new geography" that emphasizes analysis, explanation, and significance.

What is the capital city of Nigeria?

To highlight the difference between old and new geography, first consider this question: What is the capital city of Nigeria? Do you know? The question is classic "old geography," and the answer is Abuja.

Why is Abuja the capital of Nigeria?

Now consider this question: Why is Abuja the capital city of Nigeria? That's right, "Why?" This question is classic "new geography" because it involves analysis, explanation, and significance. The capital of Nigeria could be any number of cities. Indeed, until 1991, the capital was Lagos. A country doesn't just decide to move its capital every day. So why did the Nigerians move theirs? Here are three reasons:

- **A pleasant setting for expansion:** Lagos occupies a low-lying peninsula. It has little room for expansion, and the climate is hot and muggy. Abuja has plenty of room for expansion and, being located in the Central Highlands, has a climate that is much more pleasant.

- **In the middle of it all:** Lagos is on the fringe of the country. Abuja is in the middle. Having the capital in the center of the country is important because Nigeria is a developing country with a commensurate transportation system. That's a polite way to say travel can be tedious and difficult. Thus, a central location maximizes access to the seat of power and has important symbolic value, too.

- **Peace and harmony:** Nigerians are divided into some 200 ethnic groups. Some are large and have a history of mutual animosity, which, exacerbated by religious differences, sometimes manifests itself in riots and killings. Ethnically and culturally, therefore, Nigeria is something of a powder keg. So government planners sought to locate the capital in an area that is not dominated by any of the big ethnic groups nor by a single religion. Abuja fit the bill.

To sum up, I asked two questions: "What is the capital of Nigeria?" and "Why is Abuja the capital of Nigeria?" Nothing is wrong with either question. But I trust you agree that the second is the more profound of the two. It calls for a deeper, more analytical brand of thinking. As "new geography," it leaves you with a more penetrating perspective on the geography of Nigeria and the significance of a number of factors. Chapter 2 expands on how to "think" geographically.

Getting to the Essentials

In addition to focusing on the "new geography," this volume makes use of unifying concepts that will help you to understand the breadth and structure of geography. But what are these unifying concepts? Yogi Berra once supposedly ordered a pizza pie and was asked if he wanted it cut into four slices or eight. He opted for four and explained, "I don't think I can eat eight." Whether or not the story is true, a pizza pie is a pizza pie, no matter how you slice it up. The same is true of geography. In a manner of speaking, it's a very big pizza pie. Over the years, geographers have devised different ways to cut it up in order to help people like you grasp its breadth and content.

The "geography pizza slices" I'm going to introduce you to are The Six Essential Elements. They were developed as part of the National Geography Standards (see *Geography For Life: The National Geography Standards, 1994*, pages 32-35, published by Diane Publishing Company), which describe in detail "what the geographically informed person knows and understands." The National Geography Standards serve as a guide to education reform in the United States as it pertains to the teaching of geography. They were written with the advice and input of professionals who specialize in diverse aspects of geography and, accordingly, represent a broad consensus of the scope and structure of geography. Specifically, therefore, I have chosen The Six Essential Elements to describe the content of geography for the following three reasons:

✔ They are more up-to-date than any alternative scheme and take a very broad, inclusive view of geography.

✔ As part and parcel of the National Geography Standards, they have a degree of authority and authenticity that alternative sets of unifying concepts cannot match.

✔ They are probably imbedded in your local public school curriculum. If yours is one of the majority of states that recently has undergone standards-based education reform, then schools in your area probably utilize the National Geography Standards and, thus, the six essential elements, which are at the heart of these standards. The six essential elements are:

- The world in spatial terms
- Places and regions
- Physical systems
- Human systems
- Environment and society
- Uses of geography

These may sound somewhat imposing, but rest assured, they refer to simple concepts that you encounter in your everyday life. Indeed, you are already familiar with each of them, though perhaps not by their formal titles. I can prove it to you.

Where things are in the world: The world in spatial terms

You probably have a preferred grocery store, clothing store, and restaurant, plus a map in your head that tells you where they are and how to get to them. What's more, you could probably conjure up a route to visit all three in a single excursion and draw me a sketch map of the itinerary. If so, then you are already familiar with the world in spatial terms (see Figure 1-1).

Figure 1-1: When you drive, you can't help but come to grips with the world in spatial terms.

Spatial refers to the location and distribution of things and how they interrelate. Accordingly, the world in spatial terms responds to geography's most fundamental question: Where? Getting a handle on this element involves:

- ✔ Knowing how to use and read maps and atlases
- ✔ Acquiring a general understanding of the tools and techniques that geographers use to accurately locate things
- ✔ Being able to indicate the location of something using the system of latitude and longitude, or plain language
- ✔ Seeing relationships that explain the locations of things
- ✔ Recalling from memory the location of things on Earth's surface

These are basic skills to build on. On top of that, you'll never have to worry if somebody tells you to "Get lost!"

Chapter 2, which shows you how to think like a geographer, is very much about understanding the world in spatial terms. Chapters 3, 4, and 5 are devoted to location and maps, and, therefore, focus rather directly on this element. In addition, most other chapters will contain at least one map. Thus, you will encounter the world in spatial terms again and again throughout this book.

What locations are like: Places and regions

What'll it be for your next vacation? The mountains? The shore? Chances are you have mulled over questions like these that concern different areas with different characteristics. If so, then you are already familiar with places and regions.

Place: What a location looks like

Place responds to another important geographical question: "What is it like?" *Place* refers to the human and physical features that characterize different parts of Earth and that are responsible for making one location look different from the next. The terminology may puzzle you, because in everyday speech, people commonly use location and place interchangeably. In geography, however, these two terms have separate and distinct meanings. *Location* tells you *where. Place* tells you *what it's like.*

Region: A bunch of locations with something in common

A *region* is an area of Earth, large or small, that has one or more things in common. So when you say "I'm going to the mountains" or "I'm heading for the shore," you refer to an area — a region — that has a certain set of characteristics over a broad area. Figure 1-2 shows a sandy region.

Figure 1-2:
This sandy place is character-istic of the region known as the Sahara Desert.

Regions make it easier to comprehend our Earthly home. After all, Earth consists of gazillions of locations, each of which has its own particular and peculiar characteristics. Knowing every last one of them would be impossible. But we can simplify the challenge by grouping together contiguous locations that have one or more things in common — Gobi Desert, Islamic realm, tropical rainforest, Chinatown, the Great Lakes, suburbia — Each of these is a region. Some are big and some are small. Some refer to physical characteristics. Some refer to human characteristics. Some do both. But each facilitates the task of understanding the world.

Features that characterize different locations on Earth and, therefore, epitomize the concept of place, will be the focus of several chapters. These include landforms (Chapters 6 and 7), climates (Chapter 10), population (Chapter 11), culture (Chapter 13), economic activity (Chapter 15), and urbanization (Chapter 17). Each of these characteristics, of course, pertains not only to particular locations, but also to large areas as well. Thus, they also serve to characterize and define regions.

Why things are the way they are: Physical systems

I bet you have a favorite time of year, a favorite season. You probably also have a least-favorite season. No doubt you can tell me why you like some seasons more than others, and you can probably sprinkle your rationale with personal anecdotes about good times and bad. If that sounds about right, then you are already familiar with physical systems. Figure 1-3 shows one type of physical system.

Figure 1-3: A tropical rainforest is a physical system that is produced and maintained by specific climatic characteristics.

Atmosphere, land, and water are the principal components of the physical world. Geography seeks to understand how these phenomena vary from one location to the next and why. Geographers aren't content to know what the world looks like. They also want to know how it works. That involves understanding the natural processes that shape and modify Earth's surface (see Chapters 6 and 7), cause particular climates to occur in particular places (see Chapters 9 and 10), or result in some parts of Earth having too little water and others too much (see Chapter 8).

Giving that human touch: Human systems

Have you ever visited a locale that has many more or many fewer people than where you live? Have you ever moved a long distance? Have you ever visited a foreign country? Have you ever noticed that most of your shoes and clothing are made in a foreign country? Have you ever attempted to talk to someone, only to discover that person does not speak your language? If so, then you are already familiar with human systems. Figure 1-4 shows an example of the human system.

Human beings characterize Earth's surface. That is, not only do humans live here, but by constructing cities, making farms, laying railways, and building other things, humans are an actual part of Earth's surface. Culture, trade, commerce, and government largely determine the specific ways in which people are part of the Earth. And because these institutions are so diverse, so, too, are the human characteristics that are part of Earth's surface. Key aspects of human geography will be dealt with in separate chapters. They include population characteristics (see Chapter 11), movement and migration (see Chapter 12), cultural attributes (see Chapter 13), division of Earth into political units (see Chapter 14), economic activity (see Chapter 15), and urbanization (see Chapter 17).

Figure 1-4:
The culture of these Arabs is a human system that differentiates their part of the world from others.

Interacting with the world around us: Environment and society

Do you remember a farm or piece of countryside that is now a shopping center or a housing development? Have you ever experienced air pollution or water pollution? Have you ever had to cope with a severe storm, flood, or earthquake (see Figure 1-5)? If so, then you are already familiar with environment and society.

Figure 1-5:
The effect of this hurricane on human habitation rather powerfully demonstrates the relationship between environment and society.

Human beings and the natural environment interact in many ways. For example, people play a very important role in shaping and modifying the natural world. Some results of this interaction may be visually pleasing, such as the skyline of Paris, or the terraced rice paddies of Southeast Asia, or the English countryside. But other results may be troubling, such as pollution and global deforestation. References to human impact on the environment will appear in several chapters, particularly the ones dealing with water (see Chapter 8), natural resources (see Chapter 16), and urbanization (see Chapter 17). Most importantly, an entire chapter will be devoted to matters of environmental quality (see Chapter 18).

And while people impact the environment, environmental phenomena impact people. Climate affects agriculture and other human activity (see Chapters 9 and 10). Landforms and related processes and hazards affect life and property (see Chapters 6 and 7). The geography of water impacts settlement and commerce (see Chapter 8). In a nutshell, relationships between environment and society are pervasive and profound — and for those reasons will manifest themselves in several chapters.

Putting geography to use: Uses of geography

Have you ever used a road map to plan a trip? Have you ever visited a historical site and looked at maps and exhibits that help you understand the past? Have you ever attended a meeting or read an article concerning a proposal that would change the physical character of your neighborhood? If so, then you are already familiar with the uses of geography.

You can use geography to understand the past, interpret the present, or plan for the future. That is, you can use geography to understand the extent of former empires, to understand why your city looks the way it does, or to choose the location of a new factory. Geography is, therefore, a very useful and powerful tool. To help reinforce this point, every one of the content chapters (see Chapters 2 through 18) will contain specific examples of putting geography to practical use. In addition, the concluding Part of Tens contains a chapter on careers in geography (see Chapter 21).

Chapter 2

Thinking Like a Geographer

In This Chapter

▶ Thinking geographically

▶ Taking a look at two case studies

Geography is as much a way of thinking about the world as it is a body of information and concepts. Therefore, if you want to become good at geography, you must learn to think geographically. Remember when you were in the third grade and the teacher said, "Let's all put on our thinking caps"? Cute line, wasn't it? Well, I'm asking you to put on your thinking cap — your geography thinking cap, that is.

Thinking geographically is a process that involves a discreet set of skills. Therefore, this chapter is very different from the rest because it's not, on the whole or in part, about the content of geography. Certainly, you will encounter a fair amount of information about a particular part of the world. If you remembered it, great, but that's not the point. Instead, the goal is for you to learn how to think geographically and see that doing so facilitates a deeper understanding of the human and natural phenomena that geographers study.

Changing the Way You Think — Geographically

In Chapter 1, the content of geography was likened to a pizza pie, and The Six Essential Elements were presented as a way to "cut it up." The same National Geography Standards that give us those Elements also present a series of related skills that together constitute the process of thinking geographically. They include:

✔ **Asking Geographic Questions:** Thinking geographically typically begins with the questions "Where?" and "Why?" Sticking with pizza, one might want to know where all of the pizza shops in town are located and why they are there. Conversely, a person going into the pizza business may want to know where a good location would be to open a new pizza shop, and why.

✔ **Acquiring Geographic Information:** *Geographic information* is information about locations and their characteristics. If you want to know where all the pizza shops are and why, then a first step may be to consult the Yellow Pages or some other directory. You may also visit the sites and acquire information about their characteristics. Similarly, someone going into the pizza business may do the same thing in order to learn the locations and characteristics of the sites that competitors have previously chosen.

✔ **Organizing Geographic Information:** After geographic information has been collected, it needs to be organized in ways that facilitate interpretation and analysis. This may be achieved by grouping together relevant notes, or by constructing tables, diagrams, maps, or other graphics. Thus, the person who wants to understand the geography of pizza shops might produce a map of them based on information previously acquired. The person who is considering going into the pizza business may do the same.

✔ **Analyzing Geographic Information:** Acquiring and organizing geographic information paves the way for analyzing geographic information. This is when the most heavy-duty thinking occurs. Analysis involves making comparisons, seeking relationships, and looking for connections between geographic information. What factors explain the locations of existing pizza shops? What factors make for a great location for a future pizza shop? Analyzing geographic information is kind of like playing a mystery game in which you use the information you previously acquired and organized to solve a puzzle.

✔ **Answering Geographic Questions:** The process of thinking geographically culminates in the presentation of conclusions and generalizations based on the information that has been acquired, organized, and analyzed. It may reveal, for example, that pizza shops tend to be located in places that are readily accessible to a large number of people or that have lots of passers-by. Those conclusions may, of course, prove very useful to the person who wants to open a new pizza shop and is looking for the best possible location.

Thinking geographically entails two lines of thought that are similar as well as different. They are alike in that both involve the bulleted points listed previously. The difference is that one approach focuses on where things *are* located, while the other ponders where things *should be* located. To highlight this difference, the discussion above repeatedly referred to two people. One was trying to understand where pizza shops *are* located, and the other who was trying to determine where a pizza shop *should be* located. The following cases studies help reinforce these perspectives. Each poses a geographic question and challenges you to analyze geographic information before you arrive at an answer. That is, each has you thinking geographically. In doing so, you begin to acquire and develop important conceptual skills that constitute major mileposts in becoming a true geographer.

Case study #1: Where something is located

Where are African lions located and why? Obviously they live in Africa, but in what parts of Africa, and why? Those geographic questions are central to our first case study.

I'd love to be able to pack you off to Africa and have you acquire relevant geographic information, but that's not very practical. Instead, I simply refer you to Figure 2-1, which presents geographic information that has been acquired and organized in a map. So where are African lions located? What's the message of the map?

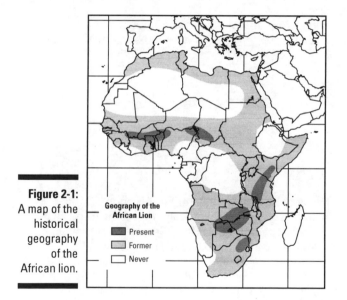

Figure 2-1:
A map of the historical geography of the African lion.

Geography of the African Lion

■ Present
▨ Former
☐ Never

The answer is that African lions are much less widespread than they used to be. The map tells you this by using three kinds of shadings, the meanings of which are shown at the lower left of the map. One shade shows areas where lions are found at present. Another depicts where lions formerly roamed. The last indicates areas where, as far as anyone can tell, lions have never lived.

A fraction of its former self

Today, African lions in the wild live only in the handful of patches shown on the map, mainly the ones in southern and eastern Africa. But the map also tells us there was a time when the lion's homeland consisted of a vast and contiguous hunk of Africa that stretched all the way from the Mediterranean coast in the north to the southernmost tip of the continent. Look at the map and visually compare the amount of territory that is lion country today versus the amount of former territory. I'm going to guess that the total land area that lions occupy today is no more than 15 to 20 percent of its former extent. In any event, present-day lion country is a fraction of its former size.

What in the world — or rather, what in Africa — happened to cause such a reduction in the size of lion country? Why did it happen? And what is the significance? I do not really expect you to have the answers at your fingertips. But take a few moments again, and this time see if you can't come up with some possible reasons as to why lions live where the map says that they do, and why lion country has decreased so substantially.

Where lions hang out

First of all, where do lions live? No, I'm not asking you for a street address; but rather, in what kind of environment do lions tend to hang out? Here are a few choices of where your average well-adjusted lion might live:

- In a forest
- In a desert
- In the mountains
- In a grassland
- Anywhere it darn well pleases

Although the last choice has considerable merit, the best response is "in a grassland." Lions generally live in grasslands. You may have known the answer because just about everybody has seen a TV wildlife documentary, which, in graphic detail, shows lions killing their next meal and then eating it. But just in case, next time you see one of those programs, concentrate on the physical setting instead of the kill. That's right, skip the build-up . . . the eyeing of the herd . . . the stalk . . . the chase . . . the cute little impala meeting its untimely end. Instead, focus your attention on the surrounding countryside, and what you are bound to see is that this life and death drama is playing out on what is essentially an extensive grassland.

What gives with grasslands?

But what gives with grasslands? Or rather, why do lions choose to inhabit grasslands? Here are a few choices as to why lions live in grasslands:

- Green is their favorite color.
- That's where those cute little impalas live.
- They got into grass while they were in college.

> ✔ They run into few trees.

> ✔ The rents are low.

Although each choice could be correct, the best response is "that's where those cute little impalas live." Lions love impalas.

Indeed, they truly love them to death. Like all wild animals, lions tend to live in places where they can find relatively abundant food to their liking. So lions hang out where impalas, zebras, wildebeests, and other animals are on the menu. Lions, of course, are *carnivores* — meat-eaters. And nearly all the animals on the menu are *herbivores* — grass-eaters. So lions prefer to live in a grassland because, as far as they are concerned, it's one big meat market.

Extinction made easy

Time to stop beating around the bush — and around the grassland, for that matter. The main message of the map is that lion country is a small fraction of its former size. And although the animal itself is not on the brink of extinction, things would appear to be headed in that direction. So what happened?

Perhaps it would be better if I personalized the question. Let's say you really have it in for the king of beasts and want to get rid them. I'm talking extinction. What is a safe, easy, and effective way to go about it? You have a couple of options:

> ✔ Shoot every last one of them

> ✔ Teach impalas self-defense

> ✔ Destroy their habitat

> ✔ Force them into early retirement

> ✔ Pack them off to Australia

Although each response has some possibilities, the best choice is "destroy their habitat." And that is indeed the main reason for the reduction of lion country from its former dimensions to its present ones and is also the reason why the lion is located where it is now.

A natural habitat can change for natural reasons or for unnatural reasons. As regards to the former, climate change is a major possibility. Natural grasslands are the result of a specific set of climatic characteristics. So if those climatic factors change, you would expect grasslands to change, too. Now, ample evidence exists of climate change in Africa. But the nature and extent of it is insufficient to explain the wholesale disappearance of grasslands over the wide area indicated on the map. So climate is not the culprit. Instead, the fault lies elsewhere and mainly takes the form of human beings.

Animal geography, Hollywood style

Movies may be responsible for more environmental *mis*information than any other source. Thus, in the world according to Hollywood, animals have a maddening tendency to show up in locations where they have no business being. Sometimes the errors are rather obscure. For example, in the nativity scene at the start of *Ben-Hur,* a Holstein calf prances by the manger. Holsteins are those dairy cattle with the black and white splotches. The problem is the Holsteins come from Schleswig-Holstein, the part of Germany that borders Denmark. Two thousand years ago, there would not have been a Holstein anywhere near Bethlehem. Like I said, sometimes the errors are rather obscure.

Then again, sometimes the errors are downright outrageous, and, in that regard, nothing beats Hollywood's treatment of the African lion. Check out just about any of the old *Tarzan* movies, *George of the Jungle,* or a host of other flicks set in a rainforest. Almost inevitably, one or more lions show up. The problem, of course, is that a lion has a whole lot less business being in a rainforest than does a Holstein in Bethlehem. Lions do not live in rainforests. Period. They never did, don't now, and never will. And the reason is simple. A lion has virtually nothing to eat in a rainforest — except maybe Tarzan.

Fewer lions? So what?

What, if anything, is the significance of the map and the story behind it? Is there any relevance? I believe so.

The pressure on natural habitats continues (and not only in Africa). Unless something is done to halt the tide, the great grasslands will continue to diminish and so, too, will the lions. Governments in affected areas are increasingly committed to heritage conservation and view protection of natural habitats and wildlife as part of that process. Thus, the average lion in the wild today lives in a national park or national game preserve. But pressure is being put on governments to open the parks to grazing and other activities that constitute "multiple use." Local officials must make choices that concern balancing the desire for conservation with the needs of citizens.

The situation is relevant to other lands, including the United States. Lions don't live in the wild in the U.S., but other animals do. And in many cases, their stories mimic the lion's. That is, they are much less widespread than they used to be. National parks and preserves have helped stem their decline and some species have been successfully re-introduced to some areas. But human population growth, coupled with pressure for land development and multiple uses, make the future uncertain. In the U.S., as in Africa, choices must be made. Looking at the locations of animals and their habitats and thinking geographically about them help clarify the issues and processes that are involved and encourage informed decision-making.

Summing up

The answer to our geographical question (Where are African lions located and why?) is that lions are located in the parts of Africa shown on Figure 2-1 mainly because of habitat reduction that is human in origin. After posing the question, we analyzed geographic information that led to the answer, after which we pondered the implications of our findings to wildlife conservation elsewhere in the world. All in all, the focus was on thinking geographically so as to understand where things (African lions) *are* located.

Case study #2: Where something should be located

Where should a gas station be located, and why? Those questions are central to our second case study.

Thinking geographically about where something *should be* located has many important and useful applications. For example, consider the occupational endeavors called *planning*. That includes urban planning, regional planning, and transportation planning, to name just three. All are intimately concerned with the question of where things *should be* located. The business world also provides lots of useful applications. Choosing a good location is often an important determinant of whether an enterprise succeeds or fails. The questions posed previously call for a business decision based on the process of thinking geographically.

In this case study, assume that you want to go into the gas station business. Therefore, your relevant geographic question is "Where should my gas station be located?"

Similar to the first case study, I'd love to have you go around town and acquire pertinent geographic information. That would include finding prospective sites for your gas station, and identifying the factors that appear to be contributing to the success of existing gas stations that clearly are doing a lot of business. The latter is important because it helps you choose the prospective site that offers the best chance for success. But that's a bit much to ask. So once again, assume that the footwork has been done, that relevant geographic information has been acquired, and that it has been organized in ways that include a map (which happens to be Figure 2-2).

The map shows two land parcels that are indicated by "A" and "B". Assume each has an identical size, an affordable price, a busy thoroughfare alongside, and that other prospective sites for your gas station have been eliminated from consideration. Your final choice with be either "A" or "B." Is one location clearly preferable?

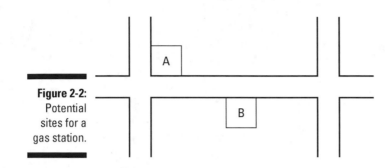

Figure 2-2:
Potential
sites for a
gas station.

Analysis of the geographical information indicates the two properties have one key difference: Property A is located on a corner lot, while Property B is in the middle of a block. Is that difference significant? Think about the location of every gas station you have ever seen. Is it on a corner or in the middle of a block? It's almost always on a corner, isn't it? And the main reason is that, on a daily basis, more cars (potential customers) pass by a corner lot as opposed to a middle-of the-block lot because the corner adjoins two roadways rather than one. In addition, corner lots are somewhat easier to enter and exit. Accordingly, the answer to your geographic question (Where should my gas station be located?) is lot A.

Summing up

In the process of choosing a location for your gas station, you have been thinking geographically once again. Only this time, however, you began by considering where something (a gas station) *should be* located. You then proceeded to acquire and organize (map) pertinent geographic information, analyze it, and answer the question.

Looking ahead

At the beginning of this case study I mentioned that thinking geographically about where something *should be* located has important applications in the fields of planning, business, and industry. Indeed, virtually every tool, concept, and content area of geography has useful applications. To reinforce this point, and to help you recognize the practical value of geography, Chapters 3 through 18 include a specific example. Be on the lookout for a sidebar whose title begins with "Applied Geography."

Chapter 3

Grid and Bear It

. . . .

. . . .

*W*elcome to Gridville, the cute little burg shown in Figure 3-1. You and I are going to pay this town a quick visit because it looks like a great locale to review basic concepts of location, plus latitude and longitude, the topics of this chapter. I say "review" because if you are like most people, then you probably learned about these things during elementary or junior-high school, but may have forgotten some or most of it later on.

Knowledge of latitude and longitude gives you basic location and orientation skills regarding our planet Earth. It also affords the opportunity to learn all sorts of little tidbits, which, in addition to impressing your friends, can greatly enhance your understanding of geography.

Feeling Kind of Square

To get started, look at Figure 3-1 and familiarize yourself with Gridville. In particular, note the following:

- ✔ The roads are aligned with the cardinal directions — that is, they run north-south or east-west. The result is a *grid* pattern of north-south roads that intersect east-west roads at right angles. So getting right with Gridville means getting used to a city that is all right angles and nothing but right angles. Thus, I'll understand if this town leaves you feeling a little square.

- ✔ North is toward the top of the map; south is toward the bottom; east is toward the right; and west is toward the left. This is a near-universal rule in map-making, but you should always carefully examine the map you are looking at and confirm which way is which.

- ✔ Gridville has a principal east-west road named Equator Boulevard, and a principal north-south road named Prime Meridian Way. The two roads cross in the middle of Gridville.

> ✔ Every other road in Gridville has a name that refers to its location relative to those two roads. Thus, streets are numbered consecutively north and south of Equator Boulevard. Avenues are numbered consecutively east and west of Prime Meridian Way.
>
> ✔ A big dot and a letter mark two intersections. I'll refer to these shortly.

Telling Someone Where to Go

Because geography involves locations and directions, it affords ample opportunity to tell someone where to go. Suppose you live in Gridville and are standing on the sidewalk at Point A, the corner of North 4th Street and East 3rd Avenue. A stranger from out of town comes up to you and asks for directions to Gridville Hospital — can you help her?

Of course you can. You know the hospital is located at Point B on the map. And you can convey that information to the stranger by stating either the hospital's relative location or absolute location.

Relative location

In the first instance, you can tell the stranger how to get to the hospital from Point A. For example (pointing west along North 4th Street), "Go that way four blocks, turn left, and walk five more blocks." This is called *relative location* because the information you gave is relative to Point A. Give those directions verbatim to the stranger at any other intersection in Gridville, and the result is a lost stranger.

Absolute location

As an alternative, you can convey the location of the hospital with respect to its *grid coordinates* — that is, its location within the grid system. For example, "Go to the corner of South 1st Street and West 1st Avenue." This is called *absolute location* because theoretically, those directions work anywhere in Gridville, not just at Point A.

The best location to use

Both relative location and absolute location have the potential of getting the stranger to the desired destination. And chances are you have used both types of location to direct someone to a destination in your town, neighborhood, or environs.

But in a global context, absolute location is far superior to relative location. When you think about it, the task of directing somebody to a location halfway around the world by means of relative location (e.g. "Go that way 11,238 miles and turn right") is rather mind-boggling. And even if you could do it, that information would only work at the one location where that information was given. It would be far better if every place on Earth had an absolute location such as that hospital in Gridville. Of course, that would be contingent on the existence of a global grid that basically mimics what we've seen in Gridville. Fortunately, such a grid exists.

The Global Grid: Hip, Hip, Hipparchus!

Like Gridville, the world as a whole possesses a grid whose coordinates may be used to identify the absolute location of things. Indeed, that is why a Greek named Hipparchus invented the global grid some 2,200 years ago.

As chief librarian at the great library in Alexandria, Egypt, Hipparchus compiled information about lands and cities all over the expanding Greek world. He saw the value of accurately locating objects on a map, but in those days that was easier said than done. Maps were notoriously inaccurate, due in good measure to lack of a systematic means of stating the location of things. So Hipparchus set out to rectify the situation, and came up with the global grid that is still in use today (see Figure 3-2).

Avoiding gridlock

Proper use of a grid coordinate system to state the absolute locations of things depends on a handful of prerequisites. Think of these as ways of avoiding gridlock:

- **Familiarity breeds success.** Knowledge of the naming and numbering of grid components is essential. If, for example, that stranger were not familiar with Gridville's grid, then telling her the hospital is at "the intersection of South 1st Street and West 1st Avenue" would have made no sense whatsoever. The same is true with respect to the global grid. That is, knowing how the lines are named and numbered is essential if you are to use the grid successfully.

- **Unique components.** Each road in Gridville and each line on the global grid must have a unique name. In Gridville, for example, there must be only one road named South 1st Street, and only one named East 1st Avenue. If multiples exist, then more than one site could satisfy "the intersection of South 1st Street and East 1st Avenue." And that would rather defeat the concept of absolute location, whether in Gridville or around the globe.

✔ **No double-crossing allowed.** Don't take that as a threat or accusation. What I mean is two roads in Gridville may cross each other only once. The same goes for two lines on the global grid. If they have multiple junctions then, such as the last point, there would be two or more intersections of, say, South 1st Street and East 1st Avenue. And again, that would defeat the concept of absolute location.

✔ **Full names, please.** You must use the full name of each road in Gridville and each line on the global grid. Again, the absolute location of the hospital is the intersection of South 1st Street and West 1st Avenue. Now suppose you had told that stranger, "The hospital's at the corner of 1st Street and 1st Avenue." Well, if you look carefully at the map of Gridville, you find four locations where a 1st Street crosses a 1st Avenue. Obviously, the potential for location confusion here defeats the purpose of absolute location. The remedy is to use the full name of each grid component.

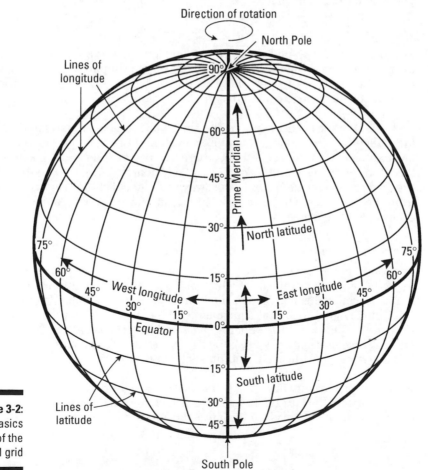

Figure 3-2:
The basics
of the
global grid

The naming game

While the Gridville grid consists of real roads, the global grid consists of imaginary lines of latitude and longitude (see Figure 3-2). *Latitude* lines go across the map — latitude comes from the Latin *latitudo,* meaning breadth, or the measure of the side-to-side dimension of a solid. *Longitude* lines run from top to bottom — longitude comes from the Latin *longitudo,* meaning length. This makes sense because when viewed on a globe, lines of longitude are generally lengthier than lines of latitude.

Similar to the roads in Gridville, the global grid contains a principal line of latitude (*the equator*) and a principal line of longitude (*the prime meridian*). All other lines of latitude and longitude are named and numbered respectively from these starting lines. It makes sense, therefore, that if you want to make like Hipparchus and draw a grid on a globe, then these are the first two lines you would draw. But where would you put them, and why?

The equator

Because Earth is sphere-like, no compelling locale cries out and says, "Use me to locate the equator!" So where to put it? Old Hipparchus might simply have said, "It's Greek to me!" and placed it anywhere. Instead, he wrestled with the challenge and came up with an ingenious solution.

He knew that the Earth is sphere-like and that it rotates around an imaginary line called the *axis.* Look on a globe and you find two fixed points, halfway around the earth from each other, where the axis intersects the Earth's surface: the *North Pole* and the *South Pole.* So Hipparchus drew a line that ran all the way around the globe and was always an equal distance (hence, equator) from the two Poles. The result is a latitudinal "starting line" from which all others could be placed on the globe.

The prime meridian

The longitudinal "starting line" is called the *prime meridian,* which signifies its importance as the line from which all other lines of longitude are numbered. Locating this line proved more problematical than locating the equator. Quite simply, no logical equivalent of the equator exists with respect to longitude. Thus, while the equator came into general use as the latitudinal starting line, mapmakers were perfectly free to draw the longitudinal starting line anywhere they pleased. And that is what they did.

Typically, mapmakers drew the prime meridian right through their country's capital city. By the late 1800s, lack of a universal prime meridian had become a real pain in the compass. International trade and commerce were growing.

Countries were claiming territory that would become colonial empires. But one country's world maps did not agree with another's, and the international climate made it increasingly advisable that they do so.

As a result, in 1884 the International Meridian Conference was convened in Washington, D.C. to promote the adoption of a common prime meridian. Out of that was born an agreement to adopt the British system of longitude as the world standard. Thus, the global grid's prime meridian passes right through the Royal Greenwich Observatory, which is in the London suburb of Greenwich, as well as parts of Europe, Africa, and the Atlantic Ocean. The British system was chosen largely because in 1884 Britain was the world's major military and economic power, and also had a fine tradition of mapmaking.

Getting Lined Up

With the starting lines in place, one can now contemplate putting all of the other lines of latitude and longitude on a globe. In doing that, Hipparchus used the notion that 360 degrees (°) are in a circle. Accordingly, he drew lines of latitude such that each and every one is separated by one degree of arc from the next. He then did the same with longitude. This is why lines of latitude and longitude are referred to as degrees.

Why is Earth 360° round?

The ancient Sumerians believed there were 360 days in a year. Like other civilizations way back when, the Sumerians equated their gods with celestial objects. Not surprisingly, the sun god was especially important. Because it took the Earth 360 days to travel around the sun (or so they believed), the Sumerians figured the number 360 had extra-special significance. As a result, they developed a system of mathematics based on multiples of 6 and 60. Nowadays, we would call it *base-6 mathematics* or (get ready for this) a *sexagesimal* system. In any event, $6 \times 60 = 360$.

The ancient Egyptians adopted the ancient Sumerians' numerical ideas, and eventually discovered the error concerning the length of the year. But by then, however, the number 360 had achieved such acceptance and status that the Egyptians decided not to mess with it. Accordingly, they kept the 360-day year but, being fun-loving people, added an annual 5-day holiday.

The ancient Greeks, like the ancient Egyptians, were adept at adopting things and ideas from civilizations more ancient than they. So when Hipparchus, in about 140 B.C., began fiddling with the notion of dividing a circle (and Earth) into degrees, he chose the number 360.

What's wrong with this map?

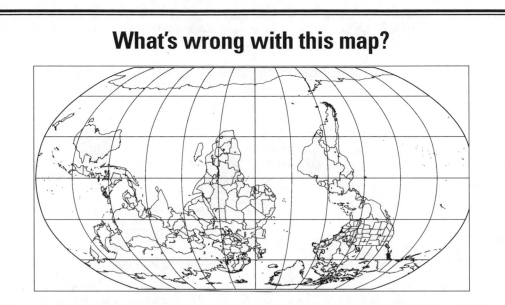

The answer to the headline is this: Nothing much, really. You could say the map is upside down, and you would be right to a point. After all, nowadays maps commonly have north at the top. But considered as a planet in the multi-dimensional vastness of space, Earth has no "right side up." Thus, no compelling scientific reason exists as to why you can't make a map with south toward the top — other than that it would look strange and confusing to most people. Indeed, in olden times maps were oriented every which way. It was only with growing availability and use of the magnetic compass in the early Middle Ages that it became common to make maps with north toward the top, just as the compass pointed. Hoping to end confusion regarding direction, in 800 A.D., Charlemagne decreed that thenceforth all French maps would be made with north at the top, that direction to be indicated by the *fleur-de-lis*. Other lands quickly followed suit. Thus, the emperor's edict became and remains the global standard.

Latitude

The system of latitude lines has the following characteristics:

- Lines of latitude run across the map (east-west) and are called *parallels* because each line of latitude is parallel to every other line of latitude.

- The equator (Latitude 0°) divides the world into the *Northern Hemisphere* and the *Southern Hemisphere*.

- Starting from the equator, each successive line (degree) of latitude is numbered consecutively both to the north and to the south as far as the North Pole (Latitude 90° North) and South Pole (Latitude 90° South).

✔ Except for the equator, each line of latitude is identified by a number between 0 and 90 and by the word North or South (or the abbreviations N or S) to indicate its location north or south of the equator. Thus, the line that is 20 degrees north of the equator is referred to as Latitude 20° North. It would be misleading and incomplete to just call this line "Latitude 20" because another line of latitude south of the equator could also be called "Latitude 20."

✔ Only one line of latitude is a *great circle,* a line that divides the Earth in half.

Longitude

The system of longitude lines has the following characteristics:

✔ Lines of longitude run from the North Pole to the South Pole (top to bottom of the map) and are called *meridians.*

✔ As opposed to latitude, no two lines of longitude are parallel to each other. Rather, successive lines of longitude are about 70 miles apart at the equator, but from there they slowly converge until they come together at the two poles (see Figure 3-2).

✔ The prime meridian (Longitude 0°) divides the world into the *Eastern Hemisphere* and the *Western Hemisphere.*

✔ Starting from the prime meridian, every line (degree) of longitude is numbered consecutively to the east and to the west half way around the world. Because Earth is 360 degrees around, 180 degrees of longitude lie east and west of the prime meridian.

✔ Every line of longitude (except the prime meridian and the 180 degree line) is identified by a number from 1 to 179, and by the words East or West (or the letters E or W) to indicate its location east or west of the prime meridian. Thus, the line that is 20 degrees east of the prime meridian is referred to as Longitude 20° East. It would be misleading to call this line "Longitude 20" because some another line that is 20 degrees west of the prime meridian also could be called "Longitude 20."

✔ Every line of longitude, is a *great circle* — a line which, if continued around the world, would divide the Earth equally in half.

Graticule

As far as geographers are concerned, latitude and longitude make for a very special grid that deserves a special name, *the graticule,* to distinguish it from every other kind of grid. Indeed, this name is so special that many dictionaries and computer spell-check programs do not recognize it. But geographers do, and they are extremely impressed if they hear it used by a layperson.

But more important than saying "graticule" is the ability to use it properly. That means, among other things, correctly identifying the grid coordinates (latitude and longitude) of locations indicated on a map. With that in mind, take a look at Figure 3-3, which represents a portion of the graticule. Note that lines of longitude are shown parallel (when in reality they converge toward the poles) and that only every tenth degree line of latitude and longitude are indicated. World maps typically "skip" lines in a similar fashion, lest they become cluttered by the graticule. But what I really want you to focus on are the three dots lettered A, B, and C. See if you correctly can identify the coordinates of each dot, keeping in mind the following rules:

1. When reporting coordinate locations, always give the latitude first, and then give the longitude. (Why? I have no idea, and I don't think anybody else does either. It's just the rule.)

2. Correct reference to latitude must specify whether a location is north or south of the equator (Latitude 0°), assuming the location is not on the equator itself.

3. Correct reference to longitude must specify whether a location is east or west of the Prime Meridian (Longitude 0°).

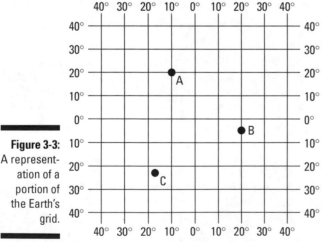

Figure 3-3:
A representation of a portion of the Earth's grid.

The correct locations of the dots are as follows:

A = Latitude 20° North, Longitude 10° West

B = Latitude 5° South, Longitude 20° East

C = Latitude 22° South, Longitude 17° West

Geo-jokes

Q. Why did the chicken cross the equator? A. To get to the other hemisphere.

Q. Why was longitude boiling mad? A. Because it had 360 degrees.

Q. Why weren't there any parallels on the map? A. Because the cartographer had no latitude in the map's design.

Q. Why were the meridians lost? A. Because they were in a parallel universe.

Had enough? I certainly have. But if this is your idea of humor, you can get more of it on the About.com geography page (`http://geography.about.com`).

Minutes and seconds that don't tick away

On Earth's surface, adjacent lines of latitude and longitude may be several miles apart, and that creates a potential problem if you wish to state the absolute location of a spot that is "between the lines." For this reason, the graticule contains a couple of levels of refinement (see Figure 3-4).

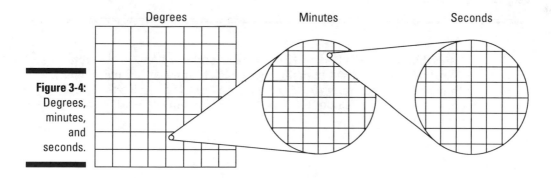

Figure 3-4: Degrees, minutes, and seconds.

Degrees   Minutes   Seconds

First, the space between successive degree lines may be subdivided into 60 equidistant units called *minutes* ('). Second, the space between successive minute lines may be subdivided into 60 equidistant units called *seconds* ("). And if more exactitude is needed, then seconds may be carried out to as many decimal points as may be necessary. I'll show you an example in just a second, but that reminds me to make a point.

Applied Geography: Using GPS to save Asia's tigers

Press a button on a GPS and it gives your precise latitude and longitude. What's a GPS? It stands for Global Positioning System and involves a hand-held device about the size of the remote control for your TV. Courtesy of the U.S. Defense Department, a number of satellites orbit Earth and constantly calculate their precise locations. When you activate a GPS, it acquires location data from 3 or 4 of those satellites and uses it to calculate its own location in degrees, minutes, and seconds of latitude and longitude. It's an amazing piece of technology that has all kinds of applications. Take saving the tiger, for instance.

Asian tigers are highly endangered in the wild because human activities (mainly farming and forestry) have reduced their natural habitat. Governments in affected areas are striving to create natural tiger preserves, but a major problem is that nobody knows exactly how much habitat a tiger needs to behave like a normal tiger. Recently, researchers were able to stun a wild tiger and attach to it a collar that contains a GPS and data recorder. Every so often the device activates, calculates the tiger's location, and records it. After a couple of weeks, the collar detaches and sends a signal that enables the researchers to find it. What they collect, of course, are data that allows them to map the animal's wanderings and therefore get a good idea of exactly how much land a tiger needs.

Doesn't this sound familiar? Sixty seconds in a minute? And for good reason. The system that you use to tell time goes back to the same Sumerian base-6 arithmetic that Hipparchus used to divide up a circle and also the world. Hmmm . . . there are 24 hours in a day. Think 24 being evenly divisible by 6 is just a coincidence? No way.

Chapter 4

Maps That Lie Flat Lie!

* *

* *

*1*magine a million-dollar map contest. The only thing you have to do to win is to supply an exact map of the entire Earth that's flat. Here's how to enter!

1. Get your hands on a globe.

2. Peel off the map of the world in such a way that you end up with one big piece of map peel. (You may want to use somebody else's globe because this procedure results in the globe's complete ruin.)

3. Lay the map peel on a flat surface so that the two surfaces are completely in contact but without distorting the original map in any way. You can cut the map if you want, but pulling and stretching it is prohibited.

You are absolutely right if you think it's going to be tough to submit a winning entry. Actually, it's impossible. You can't take a spherical surface, such as Earth, and lay it down flat without distorting the original image. This fact, however, hasn't deterred people from making flat maps of the world or parts thereof. And, to do that, the mapmaker has to figuratively *pull it here* and *stretch it there*. The result is a map that's full of distortion. Full of distortion? Well, simply put: Maps that lie flat lie!

Maps of the world are among the most basic aids to geographic learning. Many people take it for granted that they are truthful. But in reality, all flat maps of the world lie — they simply cannot help it. As a novice geographer, it is important that you appreciate that simple fact and understand the ways in which maps distort their portraits of your Earthly home. This chapter shows just how flat maps lie.

Seeing the Light: Map Projections

Accordingly, this chapter is about mapmaking with emphasis on the distortions that are inherent in flat maps of the world. But first, some basic vocabulary is in order. A *map* is a representation of all or part of Earth's surface. *Cartography* is the field of mapmaking, and a *cartographer* is a person who makes maps. Way back when, cartography was pure freehand, and I do mean way back. The oldest known map is a 5,000-year-old clay tablet that shows physical features of Mesopotamia. Later, cartography became associated with instruments and techniques that most people think of as *drafting*. Nowadays, most cartography is done with the aid of a computer.

Flat maps are called *projections* because, theoretically, making a map of the world or a large part of it involves projecting a globe onto a piece of paper or similar flat surface. Imagine, as shown in Figure 4-1, a clear plastic globe with a light source at its center. When the bulb is turned on, light passes through the glass sphere and *projects* the lines from the globe's surface onto a receiving flat surface. The result is a flat map of Earth — a projection. (Not to be a wise guy, but I really do hope you've seen the light because, theoretically at least, that is what projections are all about.)

Projection has two meanings. On the one hand, it refers to the process of transferring a globe to a flat surface. On the other hand, projection refers to the map itself, the result of the transferal. One could say, then, that projection (transferal) results in a projection (flat map).

The diagram that shows the globe and light bulb is a simple model that most people find helpful in visualizing how projections are made. In reality, projections aren't made with a glowing light bulb in the center of a globe. Instead, projections are products of mathematical formulas, trigonometric tables, and things of that ilk. The specifics are pretty tedious; fortunately, trying to explain it all in language that even I can understand is beyond the scope of this book. As a novice, it will be sufficient for you to appreciate that different projections exist, but none are totally truthful.

Figure 4-1:
Map
projection.

Earth's shape: Sphere-like, not spherical

People often say that Earth is a sphere. Not so. By definition, a *sphere* is a curved solid whose surface is always the same distance from its center, no matter at what point of the surface. Technically, Earth doesn't fit that definition. Instead, Earth is an *oblate spheroid,* meaning it is somewhat flattened at its poles, or, if you prefer, it bulges somewhat around the Equator. The average distance from Earth's center to the Equator is about 26 miles farther than the average distance from Earth's center to the poles. Compared to the size of Earth, 26 miles isn't a great distance, but it's enough to make Earth not a real sphere. It's better to say Earth is sphere-like, or an oblate spheroid.

Earth's rotation causes its oblate-ness. The speed of Earth's rotation is much faster at the Equator than near the Poles. This difference in speed may not be obvious, so think of it this way. Earth's circumference measured along the Equator is about 25,000 miles. If you stand at a spot on the Equator for one day — for one full rotation — you'll travel 25,000 miles. In contrast, if you stand a foot or two from the North Pole for one rotation, you'll only travel a few yards. Obviously, somebody who travels 25,000 miles in one day is moving much faster than somebody who travels a few yards in the same time. So, the area near the Equator is spinning much faster than other parts of Earth. The outward, or *centrifugal,* force the high speed of rotation causes is so great that Earth bulges around the Equator as a result.

Realizing Exactly How Flat Maps Lie

The business of making map projections requires a somewhat deviant personality. Cartographers know that maps that lie flat lie. They know for certain before they begin a project that it's absolutely impossible to create a flat map that looks exactly like the world. Does that deter them? Nope. No way.

Cartographers have developed literally dozens of different kinds of map projections over the years. Each one contains some degree of misinformation. If you're like most people you've given little or no thought to map projections nor have you suffered from not doing so. Or have you?

Understanding the facts about maps can't help but make you a better-informed person. Maps are a common means of communicating information. They pop up in newspaper articles, magazines, books, TV programs, and elsewhere. Because mainstream media is in the business of providing factual information, people may understandably assume that the maps they're looking at are accurate. But maps that lie flat lie, and there's nothing anybody can do about it — except maybe understand the nature of the distortions and appreciate that flat maps should be interpreted with a certain amount of caution.

There's an old saying in cartography: Close counts in horseshoes, nuclear war, and map projections. (Actually, I just now made that up; but because this chapter is all about lying, what the heck!) Cartographers know projections lie, so their objective is to get as close to reality as possible. But enough of this blabber about maps that lie, it's time to consider a practical example that involves some honest-to-goodness maps. Or rather, some not-so-honest-to-goodness maps.

Singapore, please. And step on it!

Suppose you live in New York City and are preparing for a trip to Singapore, almost halfway around the world. In planning your trip, you decide to minimize your flying time and also to stop somewhere for a day or two, just to break up your travels. A friend suggests a stopover in Rome, Italy. But another friend tells you to layover in Helsinki, Finland. You have no idea which choice is best, so you decide to find out by plotting the two cities on a map (see Figure 4-2).

Figure 4-2:
New York
City to
Singapore:
Map # 1.

Accepting the principle that a straight line is the shortest distance between two points, the map seems to make your choice pretty clear, doesn't it? The itinerary to Singapore via Rome is apparently much shorter than the route via Helsinki. As a result, you call your travel agent and make the appropriate bookings.

Upon hearing your travel plans, your second friend is shocked. "You're not going by way of Helsinki?" To show your friend the wisdom behind your choice, you take out your map and note the obvious: The linear distance from New York to Singapore is shorter via Rome. Whereupon your friend produces a map of her own (see Figure 4-3).

Figure 4-3:
New York
City to
Singapore:
Map #2.

Looking at the map in Figure 4-3, three things are suddenly obvious.

- ✔ First, the global view in this map is much different than in Figure 4-2.

- ✔ Second, the results are different, too. In Figure 4-3, going to Singapore via Helsinki appears much shorter than the route via Rome.

- ✔ Third, one of these maps is lying, but which one?

If you have a globe handy, you can determine the shorter of the two itineraries from New York City to Singapore. Get a string, pull it taut, and place it on the map so that the string connects New York City and Singapore. What you observe is that the string passes over the Arctic Circle and shows that a stopover in Helsinki is a minor detour, but a stopover in Rome is a major detour. If you don't have a globe, you can't do this demonstration, can you?

Applied Geography: Putting your best projection forward

Figures 4-2 and 4-3 provide different perspectives on air routes between New York City and Singapore. While this may seem a strictly academic exercise, airlines that compete on long-range international itineraries take the matter very seriously. There's an old saying: "Time is money." And for that reason many business travelers (if they have a choice) prefer the shortest route to get them where they're going. Airline executives know this. Accordingly, marketing strategies sometimes involve making maps that present the airline's route system in the best light possible. And doing that, of course, involves choosing the best possible projection.

Wading through lies in search of the truth

The maps in both Figure 4-2 and 4-3 are lying. But the map in Figure 4-3 provides the most accurate — that is, most globe-like — perspective regarding the shortest route between New York and Singapore. I'd really love to be able to prove that to you right here on the page of this book, but therein lies the problem — literally. This page is flat. To find out which route is shortest, you need a map that really looks like the world itself. That is, you need a globe.

Because a globe doesn't come with this book, you have to come to grips with the four ways in which maps can lie: distance, direction, shape, and area.

Most flat maps lie with respect to at least two characteristics, and some lie in all four aspects. In modest detail, here is the lowdown on exactly how and why these fibs occur.

Distance

Theoretically, transferring a curved Earth to a flat map involves selectively stretching some parts of Earth's surface more than others. For example, imagine two cities are 1,000 miles apart and the land between them gets stretched a great deal during the map-making process. Now imagine that elsewhere on Earth, two other cities are also 1,000 miles apart, but the land between them gets stretched just a little to make the very same map. On the resulting maps, the distance of 1,000 miles isn't portrayed the same.

Direction

The situation with direction is pretty much the same as with distance. By stretching a globe to make a flat map, true directions become incorrect. If some parts of the globe are stretched more than others, then a north arrow placed on one part of the map may point in a different direction than a north arrow placed elsewhere.

Actually, it's possible to make a map that keeps true directions throughout its surface. The Mercator Projection, a rather famous map introduced later in the chapter, is an example. But maintaining true direction can only be achieved by distorting something else. As the Mercator Projection shows, that something else is distance and area.

Shape

Shape refers to the outline of objects on Earth's surface. In the process of projection, you can transfer a continent or island from a globe to a flat surface while keeping its shape pretty shape intact. Then again, you can make a complete mess of things because stretching here and pulling there is part and parcel to the projection process and may play havoc with shape.

For example, compare Greenland in Figures 4-2 and 4-3. Notice that the island appears very differently in the two maps. Greenland's shape is virtually correct in Figure 4-3 because the lines of longitude meet at the North Pole, just as in reality. In Figure 4-2, however, Greenland is seriously misshapen because the lines of longitude do not meet at the North Pole but are instead spread apart in the polar area. The result is a greatly distorted Greenland.

But before we sing the praises of Figure 4-3, compare the shape of Northern Africa on both maps. Africa appears much more accurately in Figure 4-2 because in that map, the spacing of North Africa's lines of latitude and longitude are pretty much true to life. In Figure 4-3, however, North Africa appears to have become an accordion. It has been stretched laterally out of proportion to its true shape. That happens because as the lines of longitude extend outward from the center point — the North Pole — the projection excessively stretches the distance between them. As a result, North Africa has a flattened appearance.

Area

Area refers to the size of objects on Earth's surface. As is the case with shape, you can transfer (project) some features from a globe onto a flat surface while keeping sizes accurate relative to other objects on Earth's surface. Then again, you can make a complete mess of things. As to the reasons why, well, I apologize that this is sounding like a stuck record, but the simple fact is that stretching here and pulling there to make a flat map screws up the relative sizes of continents, oceans, and everything else on Earth.

Isn't there a truthful map anywhere?

Many maps are honest. But before I point some of them out to you, let me re-emphasize that flat map untruthfulness is related to Earth's curvature. Obviously, big portions of Earth involve more curvature than small portions.

A flat map of the entire world is going to lie a lot because so much curvature is involved. In contrast, a flat map of the United States has the potential of being more truthful (strictly geographically speaking) because the area of the United States has less curvature than the entire world. A flat map of the town or area in which you live — well, now we're talking little fibs as opposed to big lies because your local surroundings do not have *that* much of Earth's curved surface. And if we're talking about a map of your back yard, that could be an absolutely honest map because Earth's curvature over such a small space is virtually nil.

So, yes there are honest maps, but only ones that involve relatively small portions of Earth's surface. Geography, however, involves study of the whole Earth or portions of it that typically are bigger than your backyard. That means curvature is involved and therefore the likelihood of dealing with dishonest maps.

One and only honest map: The globe!

A *globe* is a spherical map of the world. I'm almost embarrassed to write that because everybody knows a globe when they see one. But over the years, I've been amazed at the number of people who tell me that a globe isn't a map because, according to them, maps are by definition flat. Not so. A globe is a representation of Earth; so, by definition, it most definitely is a map.

The globe is the one and only honest map of the world. Because the globe has the same shape as Earth, the appearance of Earth on a globe is free of distortion. Put differently, a globe doesn't lie flat so it doesn't lie at all. On a side note, globes are very attractive and fun to look at. Place one conspicuously in your home and guests are likely to think you have good taste and are very intellectual.

Honesty is the best policy, except . . .

Globes are truthful and the truth counts, but globes have four major disadvantages relative to flat maps.

Limited field of view

No matter how you look at a globe, you can never see the whole world at once (unless you're in a room full of mirrors, but forget that as a practical solution). Indeed, when you calculate the geometry, you cannot see even half of the world at once on a globe. However, it's often desirable to view Earth in its entirety or to visually compare far away parts of the world. These perspectives aren't possible on a globe but are possible on flat maps.

High cost

Globes are comparatively more expensive than maps. I checked the Web site of a well-known company that makes wall maps, atlases, and globes. The basic globe (12-inch diameter) sells for about six times the price of the basic wall map and about twice as much as a really good atlas. Want a map of the world without paying an arm and a leg? Buy a flat map.

Lack of detail

Because globes entail the whole world they tend to show less detail. Next time you're face-to-face with the typical desktop globe, look for the region in which you live. Unless you are a resident of a big city, there's a good chance the globe doesn't show your hometown. And suppose you wanted a detailed map of your home area. How big would a globe have to be to include that kind of information? Probably as big as the Empire State Building. Globes are good for giving you the big picture, but if you want to view an area in detail then you better get a flat map.

Inefficient data storage

Two paragraphs ago, I mentioned a globe with a 12-inch diameter. If you want to take it somewhere, you can't fold it up and put it in your pocket. It probably won't even fit in your backpack. In contrast, I have an atlas that is 12 inches long, 8 inches wide, 1.5 inches thick and contains more than 100 maps. Better still, I have a pair of CD-ROMs that contain a virtual map encyclopedia. By comparison, globes are very inefficient when it comes to data storage. (Besides, it's very difficult to walk around carrying a globe and look cool at the same time.)

How serious are these disadvantages? So serious that you'll need to amend a pearl of wisdom you learned as a kid. Honesty is the best policy except when it comes to globes. Globes are truthful, but the truth in this case comes at a very high and bulky price.

Telling the truth, but telling it slanted

It's certainly true that geography seeks to provide accurate information about Earth. It's also true that flat maps are inaccurate and therefore counterproductive to the pursuit of truth — at least in a limited sense. But the four disadvantages of globes are so serious that geographers prefer dispensing with honesty (globes) and using flat maps even though they lie. Indeed, those disadvantages of globes may be recast as advantages of flat maps:

- ✔ **Unlimited field of view:** You can show as much or as little of Earth as you want on a flat map.

- ✔ **Low cost:** Flat maps cost much less than globes. In fact, a good-sized atlas containing hundreds of maps may cost less than a single globe.

- ✔ **Accommodates detail:** Want to show a small area in great detail? Not a problem on a flat map.

- ✔ **Efficient data storage:** You can fold up a flat map and put it in your pocket. Or you can put the equivalent of a hundred globes in a single atlas and carry it in your hand or stick it in your backpack. Ever try carrying 100 globes?

The bottom line is that it's okay if flat maps lie, as long as you know you are being lied to and understand the nature of the lie.

Different Strokes for Different Folks: A World of Projections

If you are a veteran map-gawker, you know that all world maps don't look the same. And if you're not, then look again at Figures 4-2 and 4-3. Figure 4-2 looks something like a rectangle, shows the entire Earth, and is centered on the intersection of the Equator and Prime Meridian. Figure 4-3 is a circle, shows only the Northern Hemisphere, and is centered on the North Pole. As mentioned earlier, the two maps offer contrasts with respect to the ways maps lie: distance, direction, shape, and size.

The appearances in the maps differ because of different kinds of projections. That is, the maps are products of different methods of transferring the curved globe to a flat surface. Over the years, cartographers have developed literally dozens of different projections. Most maps are accurate and/or visually pleasing in some respects, although inaccurate or visually displeasing in other respects.

At this point, you may feel like saying, "Look, Charlie, why don't you spare me the details? Just tell me which projection is the best one so we can move on to the next chapter." I wish it were that simple; I really do. But the simple fact is that a winning projection doesn't exist. Every projection has good points and bad points. The trick is to know the pluses and minuses of particular projections so that choosing the best map for specific purposes is easier. It really is a case of different strokes for different folks, or at least different projections for specific situations.

If you're starting to think that this is a somewhat arcane field of study, well, you're right. As a novice geographer, you don't need to commit map projections to memory. (I know several professors of geography who don't go near this stuff!) What is important, however, is that you appreciate the variety and complexity of map projections and understand that even though all flat maps lie, some do a pretty good job of showing all or part of Earth.

All in the (map) family

Generally speaking, map projections belong to one of three families: azimuthal, cylindrical, and conic (see Figure 4-4).

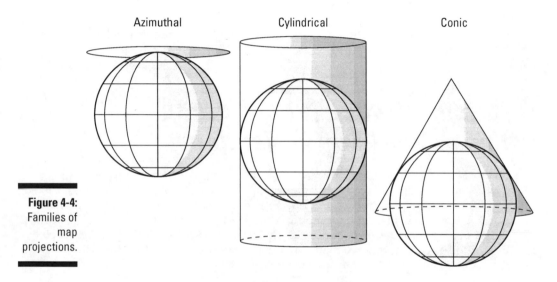

Azimuthal   Cylindrical   Conic

Figure 4-4:
Families of
map
projections.

- ✔ **Azimuthal (or planar):** A flat piece of paper (or plane, hence *planar*) is placed against the globe. The globe is then projected onto the flat paper, rendering a flat map.

- ✔ **Cylindrical:** A paper cylinder is placed over a globe. The globe is projected onto the paper. The cylinder is then cut vertically and unwrapped from the globe, yielding a flat map of the world.

- ✔ **Conical:** A conical paper *hat* is placed on the globe. The portion of the globe under the hat is projected onto the paper. The paper is cut in a straight line from its edge to the tip of the cone. The cone is then opened up and put down flat.

This reminds me to remind you that the process of projection does not literally involve projecting a globe onto a flat surface. Instead, mathematical formulas are used to plot the locations of lines (latitude, longitude, continental boundaries, and so on) on maps. Thanks to satellite imagery and high-altitude photography, you can now check the accuracy of your work in a way that was never possible before.

Five noteworthy liars

Here are five rather well known projections that represent the range of formats shown in Figure 4-4. There will not be a test over this. I repeat, there will not be a test. So don't try to memorize this stuff, but instead, just sort of let the maps visually soak in to give you an appreciation of the variety of projections that are available.

The Mercator projection

Gerhard Kremer, who's much better known by his adopted Latin name, Gerardus Mercator, developed the Mercator projection in 1569. This cylindrical projection (see Figure 4-5) is easily the most famous world map of all time. Mercator crafted his projection to aid navigation, and in that regard, the map is a gem. Straight lines on this map correspond to true compass bearings so a navigator could use it to plot an accurate course. This achievement was a very big deal in the late 16th century, and by the middle of the 17th century, a majority of Western European navigators swore by this map.

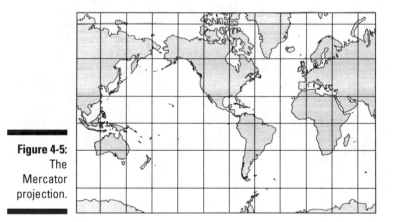

Figure 4-5:
The
Mercator
projection.

Because of its seafaring fame, the Mercator Projection later came into widespread use as a general-purpose map. That is, it found its way into classrooms as wall maps and into books and atlases. It became more or less *the* official world map, which is unfortunate because, although the shapes of landmasses are fairly accurate, the projection is extremely distorted with respect to size.

Notice that the lines of longitude on the Mercator projection don't meet at the Poles, as is the case in reality. Instead, the map shows the lines of longitude as parallel lines. This means that the North and South Polar regions have been stretched and become lines (the top and bottom borders of the map) that are as long as the Equator — 25,000 miles. One result is that land areas become disproportionately enlarged the closer they are to the areas of maximum distortion — the Poles. Alaska and Greenland are good examples. Alaska appears much larger than Mexico, while Greenland appears much larger than the Arabian Peninsula. In reality, Mexico is larger than Alaska, and the Arabian Peninsula is bigger than Greenland, but you'd never know by looking at the Mercator projection.

Because of distortions like these, the Mercator projection has fallen out of favor as a general-purpose map. No single map has replaced it. Instead, makers of wall maps and atlases have been using a number of other projections (some of which are mentioned below) that give a truer view of the relative sizes of Earth's feature.

The Goode's Interrupted Homolosine projection

Noted American cartographer Dr. J. Paul Goode (1862-1932) developed this cylindrical projection (see Figure 4-6). It's an *equal area projection,* which means that the land areas are shown in their true sizes relative to each other. In that respect, Goode's projection is far superior to Mercator's. *Interrupted* refers to the map's outline. Earth is *cut into* once above the Equator and three times below it. Therefore, the Northern Hemisphere appears as two lobes and the Southern Hemisphere as four lobes.

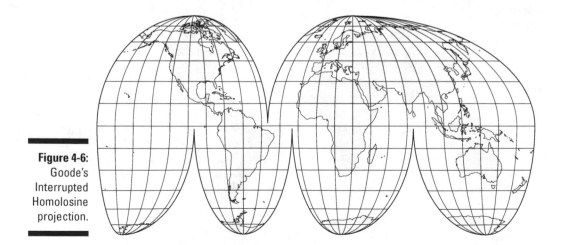

Figure 4-6:
Goode's
Interrupted
Homolosine
projection.

Why is an atlas called an atlas?

An *atlas* is a book of maps. For the longest time, maps were published singly and tended to be stored as rolled-up scrolls standing in a corner or stuck into honeycombed shelves. Gerardus Mercator was apparently the first person to compile a book of maps. Whatever the reason, his publisher decided to decorate the cover with a likeness of Atlas, the legendary Greek giant who supported the heavens on his shoulders. But in this rendering, a big globe replaced the heavens giving us the familiar image of a bent-over Atlas bearing his Earthly burden. Other books of maps copied Mercator's idea and the image of Atlas on the cover or title page became standard — which is why such volumes are called atlases. But this short history leads to speculation about what we would now call a book of maps had Mercator's publisher decided to put something else on the cover. Who knows? Maybe I'd be advising you to go out and buy a really nice aardvark.

As a result, the map's outline is not a rectangle or some other compact form, but instead is *interrupted.* The word *homolosine* reflects the fact that Goode's map is a combination of two other projections: the Mollweide homolographic and the Sinusoidal. (Whether or not you ever learn what that means, I will be happy to give you extra-credit for correct spellings.) Although Goode's projection appears in various atlases and despite its desirable equal-area attribute, many people are visually uncomfortable with its interrupted format.

The Robinson projection

Dr. Arthur H. Robinson, a noted American cartographer, introduced this cylindrical projection in 1963 (see Figure 4-2). If you lie really well, people may not notice. In fact, they may love you because of it. With all due respect and admiration to the good doctor, his map lies really well!

Although the projection contains distortion with respect to size and shape of land areas as well as to distance and direction, it has good overall balance with respect to these elements. In particular, the high latitude land areas are much less distorted than in the Mercator projection. Furthermore, Robinson's format does not have the interruptions of Goode's map. As a result of these pluses, in 1988 the National Geographic Society adopted the Robinson projection for its world maps. Partly because of the prestige and publicity of that designation, the Robinson projection has become one of the popular choices among publishers of atlases and classroom wall maps.

The Lambert Conformal Conic projection

Johann Heinrich Lambert (1728-1777), a noted German physicist and mathematician, developed the Lambert Conformal Conic projection in 1772 (see Figure 4-7). Projections cannot correctly show the shapes of large areas, but they can be drafted such that the shapes of small areas closely *conform* to reality. That is what the Lambert Conformal Conic Projection achieves.

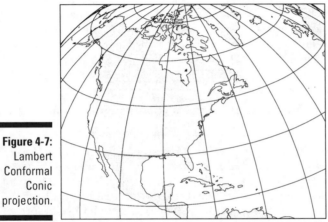

Figure 4-7: Lambert Conformal Conic projection.

Accuracy of shape (conformality) is most closely achieved where the cone, which is intrinsic to a conic projection, touches the globe. If you refer back to Figure 4-4, you can see that the conic projection makes contact in the latitudinal vicinity of the United States. For Americans, therefore, this projection is noteworthy because it is commonly used to make maps of their country.

The Lambert Azimuthal Equal Area projection

The same Herr Lambert who developed the conformal conic projection (see the preceding section) presented the Lambert Azimuthal Equal Area projection in 1772 (see Figure 4-3). Because it's an azimuthal projection, as shown in Figure 4-4, it portrays only a hemisphere, as opposed to the entire world. On the other hand, it has two positive aspects: Areas are shown in true proportion to the same areas on Earth and, as revealed in my New York-to-Singapore exercise (see "Singapore, please. And step on it" earlier in this chapter), long-range directions are depicted with a fair amount of accuracy.

Mapping a Cartographic Controversy!

If you're under the impression that the world of map-making is rather staid and geeky, you're right. In recent years, however, a map known as the Peter's projection has come along and stirred things up. Although this projection is controversial, it serves as an excellent example of why average citizens and novice geographers ought to know the facts about flat maps.

The Peter's projection was introduced and subsequently promoted in 1972 by Arno Peters, a German historian (see Figure 4-8). It's also the subject of his book *The New Cartography (Die Neue Kartographie)*. As far as accurately showing the world is concerned, this map lies with the worst of them. The appearance of the continents has been likened to wet laundry hanging out to dry. Given its distortion of the shape of land and water bodies, geographers tend not to take this map very seriously. But this projection has been adopted — even championed — by a number of influential agencies that, like Herr Peters, are actively promoting its use.

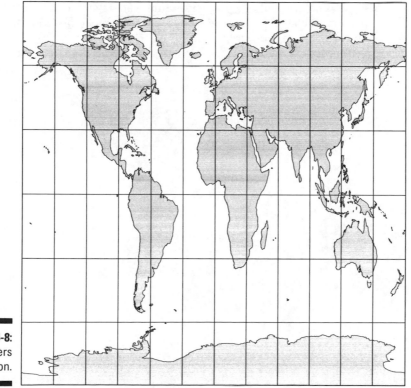

Figure 4-8:
The Peters
projection.

Advocates of the Peter's projection say it renders an important measure of cartographic justice for tropical Third World regions. They claim that by inflating the size of high-latitude regions relative to the tropics, the Mercator and some other projections present a Europe-centered view of the world that denigrates Third World countries in Asia, Africa, and South America. Proponents point out that the Peters projection is an *equal area projection* that shows tropical regions in their true size relative to, say, Europe and North America. As a result of such advocacy, several agencies with strong interests in the Third World (including the National Council of Churches, UNICEF, and UNESCO) have adopted the Peters projection as their official depiction of the world.

When you look at the facts of the matter, three things are obvious.

- ✔ First, the Peters projection terribly distorts those parts of the world it supposedly promotes.

- ✔ Second, there is a perfectly good alternative to the Peters that is an equal area map *and* depicts shape of tropical Third World regions with considerable accuracy — the Goode's projection.

- ✔ Third, there is nothing new about *The New Cartography*. The Peters projection is a knock-off of a projection that was developed by James Gall in 1885 and quickly disappeared from the radar screen of serious cartography, probably because it lies so badly.

Chapter 5

Getting the Message of Maps

· ·

· ·

*W*hen I was an adolescent, one of my prized possessions was a big world atlas. I'd sit in my room for hours at a time just pouring over pages I had randomly turned to. While it was all very mesmerizing and fascinating, one day, a disturbing thought occurred to me: Most teen-aged boys don't do this. What's wrong with me?

Over the years, and much to my relief, I met numerous geography teachers and students who, however meekly, admitted to similar habits. Indeed, such behavior turns out to be perfectly normal for people who, whether or not they know it, have a yen for geography. No doubt, that is because the map is the most basic geographical tool.

Complementing the previous chapters on Earth's grid and the properties of projections, this chapter focuses on ways in which maps communicate information and how some of that information is obtained. Basically, this chapter is about map reading and map information for the novice. Therefore, if you are, in fact, one of those people who can stare at maps for hours, then you can probably skip this chapter. But if maps confuse you or seem overwhelming, or if you have never been taught the fundamentals of maps and map reading, then this chapter is for you. While you probably won't master all there is to know, you can familiarize yourself with enough fundamentals so that you get the message of maps.

Checking Out the Basic Map Components

The basic function of maps is to show how particular phenomena are distributed over all or part of the world. In this chapter, for example, you will see maps of Canadian cities, languages in South America, and global migration to the United States. Cartographers (mapmakers) communicate these and other kinds of information in part by incorporating into their maps a standard set of elements whose purpose is to help the map-reader get the message. They include the following:

✔ **Title:** The title conveys the subject of the map and is the first thing a map-reader should look for. Ideally, its wording is simple and accurate. If the title confuses you, then that is probably more of a comment on the cartographer's communication skills than your intelligence quotient.

✔ **Legend:** Maps commonly convey information with the aid of symbols whose meanings may be uncertain. Thus, the cartographer always provides a legend (or *key*) that contains and defines the symbols found on the map.

✔ **Scale:** Scale (described more fully in the following section) provides information about the actual size of the area shown on the map. Typically, this is achieved with a small ruler-like entry on a map that equates distance in miles and/or kilometers with measurement in inches and/or centimeters.

✔ **Orientation:** Orientation is the alignment of the map with respect to cardinal directions. Which way are north, south, east and west? The standard rule of thumb is that north is towards the top of the map, but not every layperson knows the rule and not every map follows it. Accordingly, many maps include a direction indicator, minimally a north-pointing arrow. While one is included in Figure 5-1 for the sake of example, they are not common on maps of large areas because cardinal directions may vary due to distortion (see Chapter 4). Also, if latitude and longitude lines are included, then the cardinal directions are somewhat redundant.

✔ **Grid lines:** Many maps contain a couple of labeled grid lines of latitude and longitude in order to convey the global context of the mapped area. If the cartographer has reason to believe that the map reader is intimately familiar with the mapped area, or if previous maps have indicated the global context of the mapped area, then grid lines may be omitted.

✔ **Source:** Out of courtesy and honesty, cartographers commonly provide the source of the information conveyed on the map. This is especially true for maps that portray numerical data. On the other hand, some subjects are generally considered "common knowledge" and do not require source citation. Examples include maps of countries, physical features, and climates.

> ✔ **Border and neat line:** Maps are commonly contained within a *border* that assumes a conventional geometric shape, usually a rectangle. The purpose is largely aesthetic, though in books and articles borders serve the practical function of clearly differentiating map and text. Typically, borders are bold lines. Often they frame a thinner *neat line* whose purpose is purely aesthetic.

Taking It to Scale

Scale is the relationship between a distance as measured on a map and the corresponding actual distance on Earth's surface. Calculating distance between locations and comparing the size of areas are two of the more important functions of maps.

Going the distance

The scale of a map may be stated in three rather different ways, described in the following sections. Figure 5-1 shows you what the three ways look like. Some maps include just one of them. Others include two, and still others all three. Perhaps the most important thing to remember is that every map has a single scale, but a cartographer has three ways to tell you what it is. If, therefore, a single map contains two or three of the scale-types, then each is saying the same thing, albeit in a different way.

Figure 5-1:
Three ways
to indicate
the scale of
a map.

Bar Graph:

```
0            1            2
├────────────┼────────────┤
          miles
```

Statement: One inch represents one mile

Representative Fraction: 1/63,360 or 1:63,360

Bar graph

A *bar graph* looks like a miniature ruler. But, whereas the ruler you use may show inches and millimeters, the one on the map shows miles and kilometers, (as shown in Figure 5-1). The principal virtue of the bar graph is that it provides a clear visual reference to the size of the area portrayed on the map. For actual measurement, however, it may be a bit unwieldy because you can't pick it up like you can a real ruler.

Verbal scale

A *verbal scale* (also called *statement of scale*) communicates the relationship between map distance and real-world distance in a sentence or sentence-like format. In Figure 5-1, "One inch equals one mile" is the example. If you have a foot ruler handy, apply it to the bar graph and confirm that a distance of one mile as indicated on the bar graph is, in fact, an inch in length. Again, a given map has a given scale, and therefore the different ways of expressing that scale must agree, which is precisely what your ruler should demonstrate.

As far as most people are concerned, the verbal scale is particularly convenient for measuring distances on a map, provided a ruler is available. In the case of "one inch equals one mile," one need only measure the number of inches between two points to arrive at the number of miles that separate them on Earth. If, on the other hand, the verbal scale on another map reads "one inch equals 20 miles," then the number of inches between the two points on the map needs to be multiplied by 20 to render the actual distance.

Maps come in different scales. Thus, the scale you use to calculate distance on one map may not be the same for the next map. Always check the scale before you calculate distance.

Whence comes the mile?

A mile is a unit of linear measurement that equals 5,280 feet. While most of the world has adopted metric units (kilometers), Americans continue to express distance in miles, which, therefore, commonly appear as units of measurement on maps made in the U.S. But exactly what is a mile? And why does it consist of 5,280 feet instead of a more convenient figure, like 5,000?

"Mile" comes from the Latin *milia,* meaning thousand. In Roman times, a unit of linear measure called the *milia passum,* or thousand paces, was common. Somehow, somebody's thousand strides became a standard Roman mile, equal to about 1,650 yards. This measurement became widely used in Britain following the Roman's invasion. After the Empire's demise, however, the *milia passum* fell into disuse, although "mile" endured in the British vocabulary as a word applicable to a substantial distance.

The mile's present length has its origins in medieval English agriculture. Back then, a team of oxen was used to pull a heavy wooden plow. The farmer walked behind, making liberal use of an ox goad — a big stick — to influence the animals' behavior. The stick was known as a *rod,* and at some point its length was standardized to 16.5 feet. The length of a parcel of farmland was "a furrow long," or *furlong.* That was the distance the oxen could pull the plow before the farmer had to stop and rest them. Naturally, that length varied. In time, however, the furlong was standardized to a distance of 40 rods (660 feet or 220 yards). Sometime later, a distance of 8 furlongs (5,280 feet or 1,760 yards) became the standard mile, and remains so to this day.

Representative fraction (RF)

The area shown on a map is a fraction of its actual size. Appropriately, therefore, scale may be indicated as a *representative fraction (RF)*, which states the ratio between a unit of distance on the map and the same distance measured in the same units on the ground. As far as most people are concerned, this is the most confusing scale-type and the most difficult to explain. OK, here goes.

Check out Figure 5-1 again. The RF shown is 1:63,360. That means the map is 1/63,360th the size of the area it shows. Stated differently, a distance of one inch on the map equals 63,360 inches on the Earth's surface.

Once more, a given map has a given scale, but you can express it in different ways. In the example, therefore, "One inch equals one mile" and "1:63,360" must mean the same thing. And, indeed, they do. Proof is obtained by calculating the number of inches in a mile. To do that, multiply the number of inches per foot times the number of feet per mile ($12 \times 5,280$). The answer is 63,360, so the statement of scale and the RF are, in fact, the same.

Comparing Earth at different scales

Maps come in different scales; and because they do, the amount of area and degree of detail shown on one map may be very different from another. This is demonstrated in Figure 5-2, which shows three maps that have identical dimensions and progressively "zoom in" on Chicago. Specifically:

- ✔ In Figure 5-2a, 1 inch represents 630 miles. As a result, this map shows a comparatively large area that includes most of the Great Lakes, Upper Midwest, a handful of major cities, and a portion of Canada.

- ✔ In Figure 5-2b, 1 inch represents 190 miles. What is shown now is a much smaller area that includes parts of Lake Michigan and Midwest states, a few medium-size towns, and a few major regional highways.

- ✔ In Figure 5-2c, 1 inch represents 64 miles. Now we have "zoomed in" to the extent that the map shows Greater Chicago, southern-most Lake Michigan, more municipalities, local highways, and several streets.

Notice that as the area shown on these maps decreases, the amount of detail increases. And if you think about it, that makes a great deal of sense. When 1 inch represents 630 miles — a large area — only very large surface features (such as the Great Lakes) can be shown. But when 1 inch represents 64 miles — a much smaller area — then comparatively small surface features (such as roads) can be effectively shown.

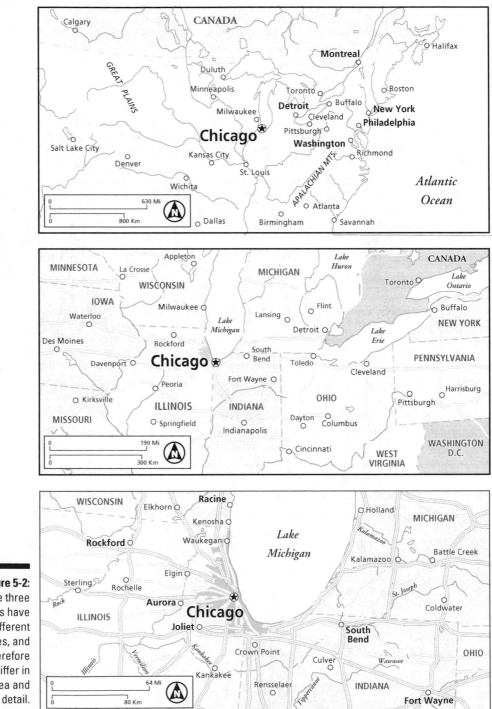

Figure 5-2:
These three
maps have
different
scales, and
therefore
differ in
area and
detail.

In the lingo of cartography, *small scale* maps show large areas in little detail, while *large scale* maps show small areas in big detail. Figure 5-2a has a comparatively small scale. In contrast, Figure 5-2b has a somewhat larger scale, while Figure 5-2c has the largest scale among the three maps. And indeed, as the scales of these maps get larger, the degree of detail increases.

Showing the Ups and Downs: Topography

All points on Earth have an elevation with respect to sea level. Altogether, they constitute "the lay of the land." (Keep in mind that elevation also pertains to points on the ocean's bottom.) *Topography* is the art and science of depicting heights and depths on a map. Like scale, topographic information is a basic feature of many maps and is commonly represented in three ways as indicated in Figure 5-3. The following sections discuss the three ways of showing topographic information.

Figure 5-3: The terrain of a make-believe place, as depicted three ways.

Spot heights

A *spot height* is a symbol (typically a tiny dot, plus sign, or triangle) accompanied by a number that indicates the elevation of a given point in feet or meters (see Figure 5-3a). Sometimes a cartographer wishes to emphasize something other than topography on a map, yet provide elevation information for a few selected points in order to convey the lay of the land. Spot heights serve this purpose.

Distortion for a purpose

A *cartogram* is a map in which different areas are distorted in proportion to numerical data. Following are two maps of Australia. The one on the left shows the true shape of the continent. The one on the right is a *cartogram* in which the sizes of Australia's states and territories are distorted in proportion to their populations. As a result, the cartogram looks much different than the "real thing." New South Wales, which is home to Sydney (the nation's largest city), contains about 10 percent of the country's territory, but about 34 percent of its population. On the cartogram, therefore, New South Wales appears bloated. In contrast, Northern Territories accounts for 17 percent of the country's territory but only 0.01 percent of its population. On the cartogram, therefore, Northern Territories is quite small. These extremes visually highlight Australia's uneven population geography. Usually, cartographers seek to minimize map distortion. In the case of cartograms, however, distortion is the purpose of the exercise.

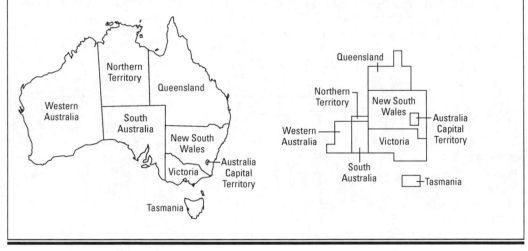

Contour lines

Contour lines connect points of equal elevation. In so doing, they convey the shape (hence, "contour") of the land they depict. Near the top of Figure 5-3b is a thin line labeled "50," which connects points that are 50 feet above sea level. Farther inland is a line labeled 100, which connects points that are 100 feet above sea level. Thus, a walk from the water's edge to a point on the second line involves a 100-foot gain in elevation.

Shading

Colors and gray tones may also be used to indicate elevation above sea level. On color maps, deep green is usually used to depict low-lying coastal land.

Light green and yellow are used for progressively higher lands, followed by light brown and dark brown. The peaks of really high mountains are often shown in white.

When gray tones are used in cartography, the general rule of thumb is "the darker the gray tone, the greater the value of whatever is being mapped." Accordingly, and as seen in Figure 5-3c, the lightest shade indicates the lowest-lying land, while deeper shades signify progressively higher elevations.

While spot heights and contour lines identify the precise elevations of precise locations, shadings refer to a range of elevations over an area. Thus, the lightest gray tone on Figure 5-3b signifies land that is anywhere between sea level and 100 feet above sea level.

But that's just this one map. On a different map, the same gray tones may mean something very different. Similarly, a light brown color may signify a particular elevation on one map, but a very different elevation on a different map. Remember: Always check the legend to make sure of the meaning of particular shades.

Using Symbols to Tell the Story

As highlighted by the discussion of topography, maps commonly show things by means of point, line, and/or area symbols. Each category, in turn, may display either qualitative or quantitative information. That is, each can simply show where something is located, or how much of something exists at a particular location or over a particular area.

Point symbols

Point symbols are used to locate discreet phenomena on Earth's surface. Most fall within one or more of the following categories.

- **Nominal icons** are tiny likenesses or symbols they name (hence, nominal) and indicate the locations of particular landscape features. Thus a tiny black dot (•) may be used to symbolize a residence while a cross (†) may be used to locate a cemetery. Whatever the symbol, the cartographer must explain its meaning in the map's legend.

- **Ordinal icons** are very much like nominal icons except that they come in different sizes that suggest comparable size or order (hence, ordinal). On some maps, for example, a tiny airplane might be used to symbolize a small airport, while a larger airplane is used to indicate a major airport. Similarly, a lower case u might be used to pinpoint a minor uranium deposit while a capital U locates a major one.

- ✔ **Dots** are often used to show how the distribution of something varies numerically from place to place. Thus, for example, a map showing the geography of dairy cattle might use a series of dots, each one representing, say, 100 head of cattle. Similarly, a map of tobacco farming might use a series of dots, each representing, say, 100 acres of land in cultivation.

- ✔ **Proportional symbols** vary in size in direct relation to numerical values. Thus, circles whose areas are proportional to population may indicate the locations and sizes of cities (Figure 5-4).

Figure 5-4: This map uses proportional circles to indicate the size of cities.

Legend

Population
○ 250,000
○ 1,000,000
○ 3,000,000

Line symbols

A number of important features on Earth's surface are linear in nature, meaning they look like lines, such as roads or railways. Likewise, migration, travel, trade, and other movements of interest to geography are basically linear phenomena that connect points. Accordingly, line symbols are common features on maps and take one of the following forms.

- ✔ **Nominal lines** note the locations of particular linear features, such as roads, railways, rivers, and borders. They may appear as solid, dashed, or embellished lines, the standard symbol for railroads being an example of the latter. Colors may also be employed. Blue lines, for example, are commonly used to indicate rivers.

- ✔ **Ordinal lines** vary in thickness or color to indicate relative importance. On many maps, for example, city, state, and country boundaries are progressively thicker so as to indicate the relative importance of the political units they mark. In Figure 5-4, the line that separates the United States and Canada is thicker than the lines that separate the states and provinces. Similarly, lines that symbolize roads often vary in thickness in proportion to the width of the highway or number of lanes.

✔ **Flow lines** indicate movement, travel or trade along a given route or between two points. On some maps, the thickness of the lines varies in direct proportion to the quantity or volume of the flow. Thus, on a map of immigration, arrows of varying widths may be used to indicate the volume of movement between sender and receiver regions (as shown in Figure 5-5).

✔ **Isolines** connect points of equal value with respect to a certain phenomenon. The contour lines shown in Figure 5-3b are an example. Similarly, daily weather maps often contain isolines that connect points with identical atmospheric pressure or the day's projected high temperature.

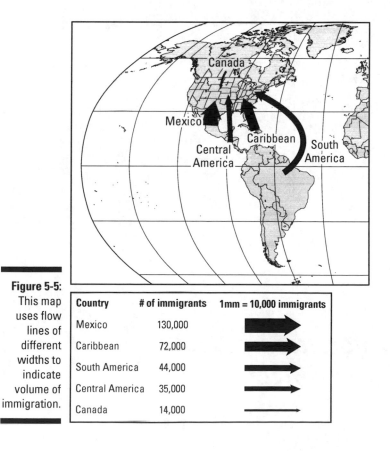

Figure 5-5: This map uses flow lines of different widths to indicate volume of immigration.

Country	# of immigrants	1mm = 10,000 immigrants
Mexico	130,000	
Caribbean	72,000	
South America	44,000	
Central America	35,000	
Canada	14,000	

Area symbols

Area symbols use gray tones or colors to depict phenomena that characterize areas as opposed to points or lines and are separated into two basic varieties.

- ✔ **Nominal symbols** identify qualitative characteristics or phenomena that pertain to areas or regions. Figure 5-6, for example, uses nominal symbols to identify official languages of South American countries. Similarly, in Chapter 10 you encounter a series of maps that use nominal area symbols to identify the geographies of the world's climate-types.

- ✔ **Choropleth maps** (from the Greek *choros* and *pleth,* meaning place and value respectively) use colors or gray tones to show how the quantity or numerical value of something varies from one area to the next. Figure 5-3c, which uses gray tones to depict elevations on an island, is an example.

Figure 5-6: This map uses nominal area symbols to identify languages of South American countries.

Official European Languages of South America

- ☐ Spanish
- ☐ Portuguese
- ▨ English
- ▨ Dutch
- ■ French

Gathering Information: Sources for Pinpointing Objects

Few things are more important in cartography than the positional accuracy of mapped objects. Historically, this was accomplished by field observation. That is, explorers or surveyors would travel to a particular area, observe locally important features, and map their locations.

Nowadays, GPS technology has greatly contributed to positional accuracy (see Chapter 3). In addition, many maps and the things that they show are products of *remote sensing.* In techno-speak, this refers to gathering information from afar about the Earth. In everyday English, it refers to use of aircraft and satellites to take pictures and picture-like images of Earth.

Aerial photography

Aerial photography refers to photos of Earth's surface taken from aircraft. Today a majority of maps produced under government approval at all levels, municipal through federal, are directly derived from aerial photography. Black and white film has been a widely used medium (see Figure 5-7). In addition to being inexpensive, it generally provides a clearer view of Earth's surface than does color film, and therefore makes it easier to identify and map surface features.

Infrared photography is also very popular. Infrared energy is contained in the sunlight that strikes the Earth and reflects off its surface. You and I can't see it, but special kinds of films and sensors can. Infrared energy readily passes through haze and air pollution, resulting in crisp images even on days when the atmosphere is far from clean. Because of that very desirable characteristic, infrared photography is widely used in aerial surveys.

Figure 5-7: This aerial photograph shows a suburban area on Long Island, New York.

The gray tones and colors on an infrared photograph may be very different than those observed on regular black and white or color film. Because of that, the term *false color* is widely applied to infrared film and photographs. Most bizarrely, vegetation appears red. Indeed, the more lush or healthy the vegetation, which appears downright green to you and me, the redder it appears on an infrared photo. Infrared photos are capable of providing information that may not be apparent on normal color or black and white photos. For example, differences in redness may indicate different kinds of crops or forests, or indicate plant life that is stressed because of disease or drought.

Non-photographic imagery

Like infrared photography, other remote sensing technologies record surface features in ways that are beyond the capabilities of human eyesight and normal cameras and film. Virtually all of them make use of sensors that scan the Earth and record surface information electronically. Because they do not use film, the pictures they produce are not, technically speaking, photographs. Thus, in the lingo of remote sensing, you have aerial *photographs* and non-photographic *images*. Three image-types are widely used.

- **Radar imaging:** In radar imaging, a sensor emits continuous beams of energy that bounce off Earth and return to the sensor, which records them. Because the emitted beams travel at a known and uniform speed, the time that it takes them to make the round trip is a function of the elevations of the locations where the beams reflect. For example, a beam that bounces off a mountaintop takes less time to return to the sensor than one that reflects off a valley bottom. This information can be used to produce detailed images of terrain and very exact topographic maps.

    Radar beams can penetrate clouds and fog with no loss of strength. Thus, radar imaging is extremely useful for monitoring and mapping Earth's surface in regions where atmospheric characteristics inhibit aerial photography (such as characteristically cloudy equatorial areas). It may also be used at night to the same effect as day. The same, of course, cannot be said of regular film.

- **Infrared imaging:** Infrared pictures of Earth may be obtained from scanners as well as films. Other than a different way of receiving images, the basic characteristics and use of infrared imaging is the same as for infrared films, discussed previously.

- **Thermal imaging:** Thermal scanners (a form of infrared imaging) record heat differences on Earth's surface. This is particularly useful for mapping ocean surface currents (whose temperature variations have a major effect on weather and climate) as well as for identifying and mapping different kinds of pollution. It has also proved very useful in mapping and monitoring forest fires and other fire-related phenomena, especially in situations in which smoke prohibits analysis by means of standard photography.

Applied Geography: The September 11th aftermath

In the aftermath of the terrorist attack on the World Trade Center, fires beneath the rubble posed significant problems for relief and rescue workers. In addition to the smoke and fumes, ongoing combustion progressively destabilized the huge debris piles in different places, heightening the danger to people on the scene. Thus, it was a matter of some importance to map the locations and intensity of "hot spots." As a result, within 48 hours of the incident, a special aerial survey was conducted that included thermal imaging. The resulting data were used to produce maps that helped on-site commanders decide where to concentrate and where not to concentrate their personnel and fire-fighting assets.

Satellite monitoring

Numerous satellites monitor and provide map-ready information about Earth's surface and atmosphere. Virtually all of them utilize non-photographic scanners that produce thermal, infrared, or radar imagery. Data received by the scanners are stored onboard the satellite for later transmission to receiver stations on the ground. There the information is processed and assembled into photo-like images (see Figure 5-8).

Figure 5-8: This infrared image shows the extent of irrigated croplands in a portion of the America West.

Getting Computerized

Today nearly all cartography at the professional level is done on a computer. Maps in this book are examples. Special kinds of software are available that allow cartographers to make maps with a degree of speed, accuracy, and data management that were unimaginable a few years ago. These qualities have also served to make mapmaking a powerful tool for a variety of businesses and planners. And in that regard, the most significant, cutting-edge field in contemporary cartography is the *geographical information system (GIS)*.

Giving you the complete lowdown on GIS would involve a lot of techo-babble that you don't want to read and I don't want to write. So perhaps the best way for me to describe GIS begins with a description of what it has replaced.

If you had poked around a city or regional planning office 20 years ago, you'd be sure to find a huge table someplace with a huge base map that showed the streets and roads of the city or region in question. There would also be numerous overlays of different phenomena drafted on individual pieces of transparent film. For example, one transparent overlay might show the location of property boundaries. Others might show land use, sewage pipes, water mains, building characteristics, telephone lines, school districts, voting precincts, contour lines, wooded areas, and anything else that may be deemed useful for planning purposes.

Again, each characteristic would be on its own piece of transparent film — that is, its own map. So if a planner wanted to see how two phenomena coincided geographically, the respective transparent films would be manually overlain on the base map and comparisons manually noted. Of course, the landscape changes. Thus, every so often a particular overlay would have to be manually updated or manually redrafted from scratch. If all of this sounds a bit tedious, then you get the point.

With the advent of GIS, all of those physical base maps and pieces of transparent film have been replaced by layers of information that exist in computer memory. This permits multiple layers, or even parts of layers, to be compared electronically, which is to say instantaneously. But the bottom line is that GIS has given geographers and planners the power to map and compare phenomena with great speed and accuracy. Indeed, remotely-sensed images can be directly "fed into" a GIS, reducing to minutes and seconds a process of field observation and mapping that used to take weeks and months.

Does this electronic gadgetry mean that the romance and adventure of maps are gone? Not necessarily. Nearly everyday I see students gawking at maps, just as I did. True, the maps they are staring at are on a computer screen instead of the pages of a book, and are more likely to be products of remote sensing instead of expedition. But the intense fascination on peoples' faces is palpable, so the old magic must still be there — the same old symbols certainly are — gray tones, different colors, tiny airplanes, and crosses. Some things don't seem to change.

Part II
Getting Physical: Land, Water, and Air

The 5th Wave
By Rich Tennant

First stop on Bob's See-the-Geographical-Wonders-of-the-World Tour is a photo op in front of a 100-foot tsunami.

In this part . . .

The natural world is big and complex. People want to make sense of it. That, as it turns out, is one of the principal tasks of geography.

Earth's surface is a mosaic of landforms covered by a rich variety of natural vegetation that is produced by diverse climate-types. Complementing this is a world of water, most of which is out of sight (and usually out of mind as well), on which life as we know it depends. None of these phenomena "just happen." Instead, mountains, plains, forests, climates, precipitation, oceans . . . every aspect of the physical world is the result of one or more natural processes that help explain the world we see and live in.

In this part, you will learn the key concepts and concerns of physical geography, which describes and analyzes the distribution of natural phenomena over Earth's surface. Yes, Earth is a big complex world. And as you will see, physical geography makes sense of it.

Chapter 6

Taking Shape: The Land We Live On

*E*arth originated about 4.7 billion years ago as a molten fireball and has been slowly cooling ever since. As a result, and after so many years, the outermost portion has hardened into a layer of rock called the *lithosphere* (from the Greek *lithos*, meaning stone). Most of this layer is so hot that it would literally fry your feet, along with the rest of you.

Fortunately, however, the outermost portion of the lithosphere is relatively cool. This sub-layer, called the *crust*, is no more than between 5 and 40 miles deep, so it accounts for a very small portion of planet Earth. But the crust has a degree of importance that is out of proportion to its volume because you live on it. The crust is your home.

The crust is also home to every kind of landform you have ever seen or will see — mountains, valleys, plateaus, plains, and so on. These features and more give character to different parts of Earth and are among the first things that come to many peoples' minds when they think about geography. And indeed *geomorphology*, the study of the nature and origins of landforms, is an important sub-field of geography.

Landforms don't just happen, however. Instead, they are products of a global war of sorts is engulfing your "crusty home." The combatants are two power-ful opposing sets of forces that shape and reshape Earth's surface. On the one hand, and the subjects of this chapter, *tectonic* forces (from the Greek *tekton*, meaning "builder") build up the Earth's crust. The pressures involved here are mighty enough to literally make mountains out of molehills, and also cause earthquakes to occur and volcanoes to erupt. Tectonic force has been modifying the crust for as long as crust has existed and will continue to do so for billions of years to come. Thus, I can say with complete confidence that the force will be with you always.

On the other hand, *gradational* forces wear down the crust. Given enough time, they can transform today's mountains into tomorrow's molehills. Gradational forces are the subjects of Chapter 7.

Starting at the Bottom: Inside Earth

The source of power for tectonic force lies deep within the Earth. For that reason, tectonic forces are sometimes called *endogenous* forces. This comes from the Greek *endon*, meaning within, and another old word that is the source of "genesis." So endogenous forces have their genesis, or origin, within the Earth.

It would be great if you could go there and see what's going on, but that's impossible. The average distance from Earth's surface to the center is 3,960 miles, and no human has ever come close. Several books and movies have portrayed such fanciful feats, but the truth is that people have barely penetrated the crust. Miners in South Africa have gone down about two miles, and if that's not the record, then the real one can't be much farther. So instead of going on a fantastic journey, you must settle for a diagram (as shown in

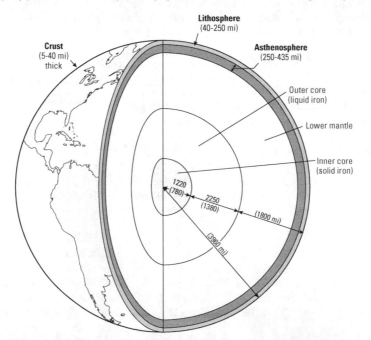

Figure 6-1:
A cut-away view of the Earth.

How do we know what's down there?

Our understanding of Earth's interior rests on a combination of inference, analysis of alien objects, sound waves, and rocks and minerals. By alien objects, I do not mean UFOs, but instead meteorites and such that have fallen to Earth. These uniformly reveal a high percentage of iron. Because these alien objects are the result of the same process of planetary formation that produced Earth, the assumption is that the proportion of iron in these objects is probably about the same for planet Earth. That suggests an incredible amount of iron beneath your feet.

Earthquakes produce sound waves. Over the past several decades, *seismologists* (people who study earthquakes) have placed within the crust hundreds of "listening devices" that record and analyze sound waves made by earthquakes. Some of these waves, it turns out, have peculiar characteristics: They cannot penetrate liquids, or liquids deflect them, or they travel at different speeds through liquids and through solids with different characteristics. Analysis of the tracks and characteristics of literally hundreds of such waves, plus the previous inference concerning iron, provide much of the input for Figure 6-1. Also, geologists have studied lots of rocks and minerals that have been thrust up through the Earth's crust. Analysis of these materials reveals a relative scarcity of iron, which suggests this substance must be concentrated deep within the Earth.

Figure 6-1). Looking at it may cause you to wonder, "Well, if nobody's ever been down there, then how do you know what it looks like?" Great question! And the answer is, it's based on informed speculation. Check out the "How do we know what's down there?" sidebar for details.

The composition and temperature of Earth's interior are the reasons nobody has ever gone there and probably never will. Most of that realm is molten or almost molten. Thankfully, not only is it out of sight and out of mind, but also out of touch. Were it not for the insulating crust, life as we know it simply would not exist.

Directly beneath the lithosphere lies the *asthenosphere*. Measured in the thousands of degrees Fahrenheit, its rock assumes a plastic, almost molten quality. Directly beneath the asthenosphere is a vast volume of somewhat stronger rock and below that liquid iron of the outer core and solid iron of the inner core that is hotter still (as shown in Figure 6-1).

Altogether, that vast volume of incredibly hot stuff is a powerful source of pressure — tectonic force. Indeed, it is mighty enough to create and rearrange continents, and in the process build mountains and cause earthquakes to occur and volcanoes to erupt. This knowledge has been available for only a couple of decades. But the idea of a force powerful enough to move continents has been around for centuries.

Moving Continents: Big Pieces of a Big Puzzle

The continents on either side of the Atlantic Ocean can easily be viewed as giant pieces of a jigsaw puzzle. If that thought has never occurred to you, then take a look at a world map, focusing on the Atlantic coastlines of Africa and Europe on the one hand, and North and South America on the other. Imagine you could place your right hand on the former, your left hand on the latter, and push the pieces together. It really does look as if they would fit together, doesn't it?

People began puzzling over this almost as soon as those shorelines were accurately mapped. Many were convinced the "fit" was too good to be mere coincidence. The implication, of course, was that there was once a super-continent that broke into big pieces that subsequently moved apart. But how could this happen? What gigantic force was responsible? And where did it come from?

Where have you gone, Gondwanaland?

Various explanations were proposed. In the 1850s, Antonio Snider suggested that during Noah's time, the Earth had several deep volcano-related cracks. Water pressure during The Flood exacerbated these cracks, created the continents, and moved them apart. A few years later, Eduard Suess proposed that a super-continent he dubbed *Gondwanaland* (after geological area in India) had fragmented and broken apart for reasons that he did not identify or endorse.

Alfred Wegener, mover and shaker

But the greatest theorizer of all was Alfred Wegener, a German geographer and meteorologist. In 1915, he proposed a theory that explained not only the shapes and locations of continents, but also the geography of mountains. According to Wegener, Earth's surface once consisted of a single super-continent called *Panagea* ("all the Earth") and a single world ocean called *Panthalassa* ("all the seas"). Pangea subsequently broke into two pieces of roughly equal size: a northern component called *Laurasia,* and a southern component called *Gondwanaland* (borrowing from Suess). Both of these, in turn, later broke up. Pieces of Laurasia became North America, Central America, Greenland, Europe, and Asia. Pieces of Gondwanaland became South America, Africa, Australia, and Antarctica.

These pieces, Wegener suggested, subsequently "drifted" apart, hence the popular name for his theory, *continental drift*. These continental "rafts" did not float on water, but instead bulldozed their way over other firmament (ocean bottom, typically). Now, at some time in your life you have probably watched a bulldozer do its thing. It scrapes the surface and produces a pile of debris as it moves forward. As Wegener saw it, that is exactly what the continental rafts were doing: scraping the Earth and producing mountains and mountain ranges along their leading edges. Thus, the "rafts" containing North and South America were drifting westerly, bulldozing as they went, the result being the chain of mountains that today extend along the west coast of the Americas from Alaska to the southern tip of South America.

Wegener's theory was basically correct, but it remained unproven until after his death because the matter of how Pangaea was broken up was unresolved. Obviously a mighty force was required to break up Pangaea and cause its pieces to move. But what could that force be, and where did it come from? Again, that was the puzzle within the puzzle.

Puzzle solved!

Imagine a mountain range about 10,000 miles long and nobody knew it existed until the latter half of the twentieth century. Sound bizarre? Well, it really happened, and of course, there's a catch of sorts. The range in question is the Mid-Atlantic Ridge. As the name implies, it pretty much runs down the middle of the Atlantic Ocean. It was virtually unknown until ocean-floor mapping (a very deep subject!) revealed its presence.

Through use of remotely controlled submersibles that carried cameras and other instruments, it was learned that this mountain range is basically a 10,000-mile long active volcano. The Mid-Atlantic Ridge is the product of *magma*, molten material from beneath the asthenosphere. A series of deep cracks, or *fissures*, in the lithosphere run the full length of the Ridge. Over the years, magma has

- ✔ oozed upwards through the fissures in response to tectonic force
- ✔ piled up on the ocean floor
- ✔ hardened to form the Mid-Atlantic Ridge

But this oozing is ongoing. Slowly, but inevitably, other magma is rising up through the fissures, to cool and harden, and in doing so, elbowing to either side of the fissure the previously hardened magma. The result is a *spreading sea floor*. The Atlantic Ocean is getting wider. The New World and the Old World are moving apart.

Subsequent deep-ocean mapping revealed other spreading sea floors in other oceans. Here was the explanation of how continents had split apart and moved! Here was the solution to the great puzzle that had been driving people nuts for years

How fast are the continents drifting?

Not very fast at all. Indeed, the proverbial "snail's pace" is much, much faster. Just as a hoot, I measured the distances between a couple of pairs of points on opposite sides of the Atlantic Ocean where the Old and New Worlds once were joined. The average distance is about 4,000 miles. Geologists say "the big breakup" occurred about 200,000,000 years ago. That means the Atlantic Ocean has widened from zero to 4,000 miles over a period of 200,000,000 years. I'll spare you my math, but it shows that the Atlantic Ocean has widened at a rate of about 1.25 inches per year. That, of course, involves two plates that are moving in opposite directions, so the speed of a single plate is about half the speed of 1.25 inches per year. (In case you are wondering, if Columbus repeated today his voyage of discovery, he would have to sail about 210 feet farther than he did in 1492.) In other words, things move slowly, but consistently. We have good evidence that the speed of plate movement varies somewhat, but at any rate, the slowest snail would leave the fastest tectonic plate far behind.

Getting Down to Theory: Earth Benedict?!

In a manner of speaking, Earth's lithosphere is like an eggshell — a thin, hard, brittle outer covering that encases a big mass of goo. Due to tectonic force, that earthly eggshell has broken into many pieces. Now, I want to be careful with this eggshell analogy. Just about everybody has dropped an egg and dealt with the mess that resulted. That won't happen to Earth. You have no cause for a nightmare in which your home planet becomes scrambled Earth, or Earth Benedict. Just remember that, like an eggshell, Earth's lithosphere is a brittle veneer that bears no resemblance to what is inside, and is easily cracked.

The Theory of Plate Tectonics explains what has been happening over the years. Here are a few points from that theory:

- ✔ Tectonic force has broken up the lithosphere into 13 large pieces, or plates.
- ✔ Tectonic force causes the plates either to move apart, collide, or slide by one another.
- ✔ Mountains, volcanoes, and earthquakes result when plates collide.
- ✔ Earthquakes may also result when plates slide by each other.

Taken together, these statements constitute much of the Theory of Plate Tectonics.

Tectonic force has broken the lithosphere into 13 plates that vary in size. The Pacific Plate, for example, covers millions of square miles. In contrast, the

Juan de Fuca Plate, which borders the Pacific Northwestern States of the U.S., is barely visible on the map (as seen in Figure 6-2). Every plate is either on the move or is being affected by the movement of a neighboring plate. Arrows on the map indicate the direction in which different plates are moving. As geographers have seen, the Mid-Atlantic Ridge marks a boundary at which neighboring plates are moving apart. Other spreading sea floors also are evident on the map. But, if plates are moving apart along some of their boundaries, that means there must be other locales where they are meeting head on, or sliding by each other. Three important results are mountains, earthquakes, and volcanoes.

Figure 6-2:
Plates
and plate
boundaries.

Making Mountains Out of Molehills

Mountains and mountainous terrain are pretty hard to miss, and everybody knows them when they see them. Likewise, most people recognize the aesthetic appeal of mountains, their value as recreational resources, and the problems they pose for surface transportation systems, land settlement, and agriculture. In addition, however, mountains are of particular interest to geography for three reasons:

- ✔ **They are climate makers.** Mountains may cause some areas to have abundant moisture and others to be bone-dry. Thus, they are major factors in the geography of climate, as you will see in Chapter 9.

- ✔ **They are culture makers.** Mountainous terrain has historically tended to isolate people and impede their ability to share ideas and material things. Thus, as you will see in Chapter 13, they have tended to encourage development of separate cultures, act as barriers between cultures, and in general, serve as major factors in the geography of culture.

✔ **They are country makers.** Because they are such visible landscape features, mountains and mountainous features — such as ridges — have often been used to designate frontiers between countries and states. Thus, as you will see in Chapter 14, they are major factors in political geography.

For now, however, the focus is on the causes and consequences of mountain building *per se* rather than the climatic, cultural, or political effects. As regards causes, mountains owe their immediate origins and locations to three processes: folding, faulting, and subduction. Each is discussed in the following sections.

Folding the crust

When plates collide head-on, a couple of outcomes are possible. One is folding, in which the crust buckles in response to the compression, and may eventually assume a rather wave-like appearance, as you can see in Figure 6-3. You can crudely simulate that sequence by doing the following: Put a piece of paper on the surface of a desk or table and place your right and left fingertips on opposite edges. Then very slowly move your hands together. Hopefully, the paper will assume a wave-like form as your fingers approach each other. Remember, however, that the Earth area represented by that paper contains hundreds of square miles, and that the convergence of your finger tips mimic plate movements that span millions of years.

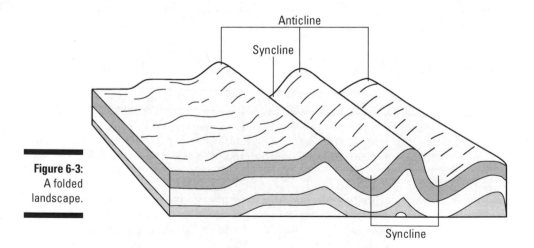

Figure 6-3:
A folded landscape.

Former grandeur

The Appalachian Mountains of the eastern United States are a prime example of a folded mountain range, as illustrated by Figure 6-4. They do not, however, coincide today with a plate boundary. This tells geographers that plates and plate boundaries come and go over the broad expanse of geologic time. In some cases, therefore, mountains mark an active plate boundary where mountain making is in progress. In other cases, mountain ranges mark ancient (extinct) plate boundaries, and are themselves mere eroded remnants of what used to be. Today, the highest peaks in the Appalachians are in the 6,000 feet range. But orientation of various rock layers suggests to geologists that in their ancient heyday, the Appalachians towered 30,000 feet and more above sea level. That's higher that Mt. Everest, the tallest mountain on Earth today.

Figure 6-4:
A photograph of exposed folds inside a hillside.

Making resources accessible

The geography of mining often coincides with the geography of mountains because the tectonic processes that make the mountains also serve to reveal the presence of valuable ores and minerals and facilitate their accessibility to humans. The coal reserves of the Appalachians are a case in point.

In the diagram of folded mountains (see Figure 6-3), assume that coal occupies the second layer, or *strata*, of rock from the surface. In the flat terrain on the left of the diagram, one can see no indication of a valuable resource underfoot. On the right, however, folding may reveal the coal. If not, then subsequent erosion — say, a river that "cuts through" the landscape — may reveal the presence of valuable strata. This, in turn, may give rise to economic activity that — such as in the case of coal mining in the Appalachians — is largely synonymous with the region.

Whose "fault" is it?

In addition to folding, head-on collisions of plates may also produce mountains by *faulting*. In this case, a series of deep fractures develop through the crust. Over time, the immense pressure attendant to the slow-motion collisions between plates may cause large-scale rock units to be raised (producing *horsts*) or lowered (producing *grabens*) along the *fault lines* that mark the intersections of the fractures and Earth's surface (as you can see in Figure 6-5). Mountains of significant size may result. And, actually, they have — the highest mountains on Earth, the Himalayan Range, are products (*fault-block mountains*) of uplift along fault lines produced by collisions between the Indo-Australian Plate and the Eurasian Plate. In this range are found Mt. Everest, the highest mountain on Earth (29,028 feet), plus K-2, Kanchenjunga, Makalu, and every one of the twenty tallest summits on Earth.

Figure 6-5: Types of faults.

As you read this, the collision between the Indo-Australian and Eurasian Plates continues. And as a result, the elevations of many Himalayan peaks are getting higher. Mt. Everest, for example, is growing by about one-half inch per year.

Plate tectonics: A four-letter word!

That word is *slow*. Mt. Everest is growing by about a half inch per year. Slow. And since the break-up of Pangea, it has taken the continents hundreds of millions of years to travel a few thousand miles. *Slow*. If you didn't read it, then please check out the earlier sidebar on the speed of continental drift. (*Slow!*) Tectonic force has massive power, but the continents and crust have massive weight. That means massive *friction of resistance* needs to be over-come if anything is going to be moved. Looked at over long periods of time, therefore, alterations to the crust due to tectonic forces tend to happen real *slowly*. Or so things normally occur. As befits a four-letter word, things can get nasty. A dark side to the force is out there.

Experiencing Earthquakes: Shake, Rattle and Roll!

The study of plate movement rather naturally leads one into the field of *seis-mology*, the study of earthquakes. These are inevitable and dangerous conse-quences of plate tectonics. Why "inevitable?" Well, an *earthquake* is a sudden movement of the Earth's crust. Given plate tectonics, earthquakes are inevitable. That is, sooner or later they're bound to happen.

During an earthquake, gazillions of tons of crust are moved. The amount of pressure required to do that is incredible; and it does not accumulate in a day. Therefore, at a given location, on an active plate boundary, earthquakes are not everyday ongoing events. Instead, it takes long periods of time — years and decades — for tectonic force to build up enough pressure to move a mass of crust.

The mechanics are crudely similar to inflating a balloon until it bursts. As the volume of air increases within, so does the pressure and tension along the surface of the balloon. Eventually, the pressure exceeds the balloon's capac-ity to contain it, and the balloon gives way . . . pop! Obviously, crust does not inflate and pop, but pressure does build-up slowly, especially along plate boundaries. Tension increases and keeps increasing over the years. And finally, perhaps after decades of pressure building, the crust just can't take it any more. And so it suddenly gives. That is, it suddenly moves, releasing the built-up pressure.

Because tectonic force exists everywhere, an earthquake can happen any-where. But given what you have read about plate tectonics, it should come as no surprise to see that the geography of earthquakes (see Figure 6-6) largely coincides with the geography of plate boundaries (see Figure 6-2). Zones of spreading sea floors are prime candidate locales. So, too, are areas where plates collide. And so, too, is another possibility that has yet to be mentioned.

Figure 6-6:
The bolded
lines show
the
geography of
earthquakes.

Splitsville in California

Sometimes neighboring plates do not diverge or collide, but rather slide by each other. The linear break in the rocks that marks the occurrence of this kind of movement is called a *transform* fault (see Figure 6-5). California's San Andreas Fault, no doubt the most famous fault line in the United States, is an example of a transform fault and is shown in Figure 6-7. The land on the western side of the fault is part of the Pacific Plate and is slowly moving to the northwest. Meanwhile, the land on the eastern side, which is part of the North American Plate, is slowly moving towards the southeast.

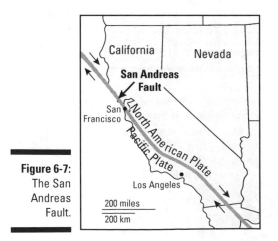

Figure 6-7:
The San
Andreas
Fault.

Actually, and as has been seen, "sudden fits and starts" is more accurate than "slowly moving." Pressure slowly builds on both sides of the fault line. Every so many years, enough pressure accumulates to overcome the friction of resistance offered by gazillions of tons of crust. At that point, parts of California slide by each other — not as continuous slow movement, but rather in short and sudden spurts as earthquakes occur. As a result, California is slowly being torn apart.

People at risk

Ultimately, geography investigates natural phenomena to gain information that is relevant to humans. Earthquakes are particularly significant because of their destructive potential and because so many earthquake-prone areas are densely populated. More important, therefore, than long-term scenarios such as the splitting of California are short-term consequences for Los Angeles, San Francisco, and several other major cities that are on or near a fault line. And California isn't the only place affected. Examination of Figure 6-4 reveals several other locations where plate boundaries coincide with major metropolitan areas. That includes about a dozen or so major cities along the West Coast of the Americas from Mexico City to Santiago, Chile. The same applies to virtually all of Japan and many parts of Southeast Asia. Then consider the long interface between the Eurasian Plate and its southern "neighbors" that extends from the Himalayas westward through Italy. All told, more than two billion people live close enough to an active fault zone to be in harm's way.

How earthquakes kill and maim

Collapsing buildings are the cause of most earthquake-related casualties. That is, most people who end up as statistics are either inside or next to a building that experiences structural failure. The walls give out. The roof caves in. The floors of a multistory building become something akin to a pile of pancakes. And people get crushed.

As a result, an adage of sorts goes, "Earthquakes don't kill people: *Buildings* kill people." Of course, those buildings didn't just up and collapse. It was the earthquake that caused the collapse, and the greater the magnitude of the quake, the greater the likelihood of casualties resulting from collapsing buildings. But rarely does a quake *per se* prove fatal. Certainly, the attendant terror has been known to induce heart attacks. But shaking earth per se is not a major killer. Usually a side effect does people in, the principal one being structural failure.

Do earthquakes "gobble up" people?

I've seen at least three movies in which the earth suddenly splits open during an earthquake and "gobbles up" people who just happen to be standing there. Yes, it has happened, but the likelihood and frequency of such occurrences is roughly once in every couple of thousand blue moons. It takes an extraordinarily powerful earthquake to make the earth "open wide" and do such a thing. Eyewitnesses to the New Madrid earthquakes, discussed in an earlier sidebar, reported seeing the earth split open. Thankfully, seismic events of such magnitude are extremely rare. Reality, however, is not always conducive to box-office success, so Hollywood has a knack for rendering the extraordinary commonplace. Powerful quakes on film that transform people into munchies are examples.

A matter of wealth and culture

Wealth and culture also play major roles in determining earthquake damage and casualties. Countries characterized by low average income and a traditional cultural environment tend to fare far worse than their wealthy, modern counterparts. Say, for example, that equally strong earthquakes strike a major American West Coast city and a city in a developing country. Chances are the toll in human lives and injuries would be far worse in the developing country.

The reason relates to differences in building construction. Buildings with walls of concrete, cinder block, brick, or adobe-like materials have a rather brittle quality, so they tend to "snap" and give way rather readily. It helps if they have a supporting skeleton of steel rods or wooden poles, but even these may prove grossly insufficient in the event of a really strong quake. Unfortunately, literally millions of buildings such as these are located in seismically suspect regions in developing countries. Each is a potential tragedy waiting to happen.

In contrast, superior construction is much more predominant in wealthier settings. Skeletal steel is much more common, as is implementation of the latest thinking regarding earthquake-resistant buildings. In that regard, an ideal model is the way a tree bends in a strong wind, absorbing the punch. The implication is to build structures that bend with the punch — or rather, sway with the earthquake. This is done by attaching girders in a way that produces a skeleton that is rather like your own — it bends.

Building codes in earthquake-prone areas of the United States, and in other parts of the world, now mandate this kind of construction. And these laws clearly are having their desired effect. But modern construction is costly — much more so than the traditional masonry that continues to dominate much of the earthquake-prone world as a matter of tradition and inability to afford the state-of-the-art alternative.

Tsunamis

When earthquakes occur under the ocean, the sea floor may rise or fall by a few feet over an area hundreds of miles on a side. This happens in a matter of seconds. With the sea surface suddenly a few feet too high or too low over a huge area, enormous volumes of water are set in motion to bring the sea surface back to level. This produces a long, low wave that moves out across the ocean in every direction at speeds that may exceed 400 miles per hour. In the open ocean the wave is less than 3 feet high, but may be 300 miles across. Although the wave is moving over 400 miles per hour, because the wave is so broad, the 3-foot rise and fall takes 10 or 20 minutes and if it passed under your ship at sea it would not even ripple the surface of your martini. You wouldn't even notice it. But as the wave encounters shallow coastal waters, it slows down and grows enormously in height, manifesting itself as a *tsunami*, or large "tidal wave." The size of the wave is directly related to the magnitude of the earthquake. Strong quakes may produce waves 30 to 50 feet in height, and 100-feet monster waves are not unknown.

Naturally, the destructive potential of these waves in regard to coastal settlements is substantial. Given the seismically and volcanically active plate boundaries that border the Pacific Ocean, more tsunamis affect that ocean's shores than any other (see Figure 6-4). Japan has been one of the most affected regions, which explains why the word "tsunami" is of Japanese origin.

When a major earthquake occurs on land, devastation is limited to a few tens or perhaps a hundred miles of the point where the quake occurred. The tsunami, however, may travel for hundreds or even thousands of miles with little loss of energy. By means of tsunamis, therefore, earthquakes can wreak havoc far away in places where the earthquake itself is never felt.

Fortunately, global earthquake monitoring now makes it possible to warn tsunami-prone cities of an approaching tidal wave, hopefully in time to evacuate people from low-lying areas. This know-how has been unevenly applied, however. Generally, affluent countries have been able to put in place good civil defense systems while poor countries have not. As is the case with earthquakes, therefore, the geography of wealth and poverty has much to do with the resulting human toll.

A matter of magnitude

The amount of destruction that results from an earthquake depends on a couple of factors, one of which is its strength. Earthquakes vary remarkably in their power. Some can barely be felt. Others can knock you off your feet or buildings off their foundations. Two methods are used to measure the power of earthquakes. Because both make use of a series of numbers, they are referred to as *scales*.

The Richter Scale

Probably the more famous of the two scales is the *Richter Scale*, which was formulated in 1935 by a *seismologist* (one who studies earthquakes) named Charles F. Richter. This scale indicates ground motion in an earthquake. It ranges from 0 to 9, but theoretically can go higher. The numbers are *logarithmic*. That means each whole number is 10 times greater than the preceding whole number. Thus, an earthquake that measures 7.0 on the Richter Scale releases 10 times more energy than one that measures 6.0, and 100 times as much as one that measures 5.0. Of course, the ground motion of earthquakes is not limited to multiples of the number 10. Thus when earthquakes are reported in Richter terms, you may see numbers such as 6.3 or 4.8.

The New Madrid Earthquake(s)

On December 16, 1811, perhaps the most powerful earthquake in the recorded history of the United States occurred near New Madrid, Missouri, located on the Mississippi River in the extreme southeastern part of that state. I say "perhaps" for two reasons. Scientific instruments that accurately measure earthquakes were not then available. And in the days and weeks that followed, two other major quakes (and literally thousands of much lesser ones) rocked the region. Either or both of those may have been stronger than the first.

How powerful were they? Apparently, each may have registered an 8.0 or higher on the Richter Scale. The Mississippi River changed course. An island in the river disappeared. New land

rose. Forests were knocked down for miles around. The ground rolled in visible waves, wiping away houses, gardens, and fields. Fortunately, remarkably few people perished, mainly because the region was then only lightly populated.

The quakes were not flukes, but instead the product of a fault zone that is minor in terms of its length, but major in respect to seismic potential. Geographically, what is most interesting is that New Madrid is more than a thousand miles from the nearest plate boundary. Thus, while the vast majority of earthquakes occur on the fringe of tectonic plates, the New Madrid episodes demonstrate that earthquakes (even very serious ones) can potentially happen anywhere.

The Mercalli Scale

Less famous is the *Mercalli Scale*. This was devised in 1902 by a gentleman whose first name was Giuseppe and whose last name you can guess. This measures the intensity or violence of an earthquake, particularly in terms of damage caused to human-built structures. Expressed by a series of Roman numerals, I to XII, the higher numbers reflect increasing damage. Table 6-1 tells you what the numbers of the Mercalli scale mean, as well as (in parentheses) their approximate Richter Scale equivalents (Note: because the Mercalli is a 12-point scale and the Richter is a 9-point scale, not every Mercalli numeral will have a corresponding Richter equivalent).

Table 6-1	The Modified Mercalli Intensity Scale
Intensity level	*Extent of Impact (approximate Richter Scale equivalent)*
I	Not felt. Usually detected only by instruments. (2 or less)
II	Felt by a few people at rest, especially on upper floors
III	Hanging objects swing. Felt quite noticeably outdoors. (3)
IV	Felt indoors by many, outdoors by few. Sensation is like a heavy truck striking a building. (4)
V	Felt by nearly everyone; sleepers awakened; trees and telephone poles shaken, some dishes and windows broken.
VI	Felt by all; books fall from shelves; glassware broken; building damage slight. (5)
VII	Difficult to stand; damage to some masonry; damage to buildings slight to moderate.
VIII	Partial collapse of masonry; chimneys fall; frame house moved on foundations. (6)
IX	Partial collapse of substantial buildings; underground pipes broken; conspicuous cracks in ground. (7)
X	Most structures destroyed; ground badly cracked; large landslides
XI	Few, if any, structures remain standing; bridges destroyed; broad cracks in ground; railroad rails greatly bent. (8+)
XII	Damage total; waves seen on ground surface; objects thrown into air.

Applied geography: Coping with tsunamis

Urban planners are using geographical knowledge and ideas to help coastal towns in tsunami-prone areas mitigate the impact of future tidal waves. Cities and towns consist of different kinds of land use — parks, apartment buildings, schools, office buildings, and so forth. A key goal of urban planning is to allocate different kinds of land use to different parts of town in ways that benefit local residents. With respect to tsunamis, land adjacent to the coastline is dangerous, but the threat lessens as one goes progressively farther inland. The typical planning response is to allocate land use such that schools and hospitals are away from the danger zone and that population density, in general, decreases with distance to the shoreline. That means, for example, placing residences, apartment complexes, and office buildings away from the shore, while allocating warehouses, open space, and other low-population density land use to the immediate coastal setting. What may sound like plain common sense is, in fact, a geography lesson that several coastal towns have learned by fatal trial and error.

Subducting Plates: Volcano Makers

Sometimes when two plates meet head-on, one overrides the other in a process called *subduction* (see Figure 6-8). The plate that gets overridden is said to be *subducting*. Only the oceanic lithosphere created at mid-ocean ridges is dense enough to *subduct* (sink) very far into the mantle below. When it does it heats up, its upper surface, or the asthenosphere, overlying it partially melts. The result is a local surplus of sorts of molten material that seeks to rise through the lithosphere, and will do so if a convenient fissure or area of weakness provides a path to the surface. When that is accomplished, the result is a volcanic eruption. Thus, the geography of subduction largely determines the geography of volcanoes.

Figure 6-8:
The process of subduction.

"The Ring of Fire"

Zones of subduction occur in many parts of the world, but are especially prevalent around the shores of the Pacific Ocean. For that reason, the Pacific Rim has an extraordinarily high concentration of active volcanoes and is known as "The Ring of Fire." Locations include the western coast of South America, the western fringe of Central America, the U.S. Pacific Northwest, much of coastal Alaska from the Anchorage area westward through the Aleutian Islands, Russia's Kamchatka Peninsula, Japan, The Philippines, and Indonesia. As is the case with earthquakes, billions of people worldwide are directly or indirectly at risk, especially so in "The Ring of Fire."

The Hawaiian "hot spot"

Hawaii is clearly a "hot spot" as regards to tourism, but that moniker applies in another way, too. Hawaii is the most volcanically active place on Earth. In most cases, subduction is responsible for volcanoes. But in a relative handful of locales, including Hawaii, the cause is a "hot spot." In these instances, hot mantle rock is rising to the base of the lithosphere and then rolling back down, like water boiling in a pot. Some of the rock melts as it rises and the magma rises close to or onto the surface. The latter is the case with the Hawaiian Islands, and is the very source of their existence.

For the last several millions of years, the Pacific Plate has been moving to the northwest, passing over a hot spot. Over the eons, magma has come up through a fissure (or vent), issued onto the ocean floor and hardened, becoming a seamount. As magma continues to seep through the vent, the seamount grows, breaks the ocean surface, and becomes an island, which continues to grow as long as the connection with the magma-giving vent remains. Eventually, however, the moving plate severs that connection and the island stops growing. A new seamount — the forerunner of a future island — then begins to grow on the ocean floor. This sequence explains the southeast-to-northwest orientation of the Hawaiian Islands, as well as why all of the islands except for the Island of Hawaii, which is now over the hot spot, are volcanically extinct. It also explains why to the southeast of the Island of Hawaii a seamount exists, which, several thousands (if not millions) of years from now, will become the next island in the Hawaiian chain.

Subduction: Another four-letter word?

Subduction is another example of that four-letter word: slow. As a result, the build-up of pressure and the basic "stuff" of an eruption are slow. Therefore, volcanic eruptions happen infrequently, as do earthquakes. And in a sense, that's the really nasty thing about volcanoes: They erupt so infrequently.

Now you may say that is a really stupid statement. Why would anybody want eruptions to happen more often? My point is this: If eruptions happened with greater frequency, then people would get the message. They would more fully realize the dangers attendant to volcanoes and be more likely to avoid the danger zones.

But that's not how it works. Eruptions happen infrequently. Subduction is slow, and there's nothing anyone can do to speed things up. So the people who occupy volcanic environs either are unaware of the dangers or like their chances. Maybe they even employ that popular, cuddly term "sleeping volcano," and pray it doesn't wake up in their lifetimes. And indeed, most days it doesn't. But then one day it does.

The big blast

A volcanic eruption is perhaps nature's most spectacular show. In some cases it consists of lava "fountains" and flows. More commonly, however, the event is a big blast accompanied by massive emissions of steam and hot rock particles of all sizes, rather than rivers of lava.

The power of eruptions is sometimes equated to many atomic bombs. Obviously, nobody wants to be around a volcano when it goes BOOM! But few people tend to live in immediate blast areas, so BOOM! *per se* is not the big thing you may think, at least in terms of immediate human casualties. In that regard, two other side effects are of greater importance: ash and lahar.

Making an ash of itself

When explosive volcanic eruptions occur, gazillions of tons of ash (tiny rock particles) are thrown into the air as a humongous, dense, and potentially suffocating cloud. This is shown in Figure 6-9. Having weight, these particles eventually fall to earth over a wide area, coating crops, covering roads and houses, and potentially causing severe (and sometimes fatal) breathing problems for people and animals. The eruption of Mt. St. Helens provides an excellent case study of the possibilities (see the "A mountain blows its top" sidebar later in the chapter).

Figure 6-9:
A photo-
graph of a
volcano in
eruption.

A mountain blows its top

I have two atlases that disagree mightily concerning the height of Mt. St. Helens. One says the summit is 9,677 feet above sea level. The other gives an elevation of 8,363 feet above sea level. That's a difference of 1,314 feet, which is close to the height of the Empire State Building. The reason for the disagreement is one atlas was published before the volcano erupted (May 18, 1980) and the other was published afterwards.

But "erupted" is a bit of a misnomer. The mountain literally blew its top. The U.S. Geological Survey estimates 3.7 billion cubic yards of mountain got blown away. Another 1.4 billion cubic yards of ash got ejected, much of it in a cloud that reached 80,000 feet within 15 minutes. Accumulations of the inevitable ash fall averaged 10 inches 10 miles downwind, 1 inch 60 miles downwind, and 1/2 inch 300 miles downwind. Fifty-seven people perished (some from the blast, others from the suffocating ash-fall), along with an estimated 7,000 big game animals and 12-million Chinook and Coho salmon fingerlings. As you can see, all kinds of numerical facts have been calculated and committed to print. What's really amazing to me is that the volume of forest that got blown down (4 billion board feet of timber) was enough to build 300,000 2-bedroom homes.

Lahars

A *lahar*, a word of Indonesian origin, is a dangerous, fast-moving mudflow. During an eruption, vast quantities of emitted steam may cool rapidly, fall as rain, mix with ash, and form a mudflow. In the cases of very high volcanoes, such flows may be complemented by large amounts of water from rapidly melted snow and glaciers. The resulting lahar may race down flanking valleys for miles and miles from the volcano, burying everything in its path.

In November 1985, the destructive potential played out to its fullest following the eruption of the Nevada del Ruiz volcano in Colombia. Lahars as deep as 150 feet raced down the side valleys. Within four hours, locations as far as 65 miles away — seemingly well beyond the volcano's reach — were under mud. Hardest hit was the town of Armero, where some 23,000 people were killed and another 5,000 injured.

In a way, a lahar is to a volcano as a tsunami is to an earthquake — a mechanism by which the power of a major tectonic event may be fatally felt far away from the actual event itself. But being far away, the possibility exists for early warning systems that can significantly lessen the number of people who end up as statistics.

Categorizing Tectonic Processes

Ironically, beautiful mountains are products of powerful processes that can kill or maim. Due to the latter, earthquakes and volcanoes are rather commonly referred to as *natural hazards* — environmental events that are potentially harmful to humans and their handiwork, such as tornadoes, hurricanes, landslides, and floods. While there is no denying the destructive potential of earthquakes and volcanoes, some students of tectonics are rather put off by the "natural hazard" name, which in their view unfairly demonizes nature and conveniently absolves humans of any responsibility for fatal effects. Better, they say, to think of tectonic events as "natural processes" that only become "natural hazards" when people get in the way, as by building houses and cities in areas at risk.

You may agree with that, or disagree with that, or have no opinion. But the statement underscores the interest of geography in this subject matter. Certainly, tectonic movements are natural processes. But when you consider the geography of these events in relation to the geography of humans and their handiwork, then tectonic forces may assume a degree of importance that goes well beyond their own immense power.

Chapter 7

Giving Earth a Facelift

*G*radational force, which wears down the Earth's crust and is the subject of this chapter, is the opposite of *tectonic force,* which builds up the Earth's crust, and was discussed in Chapter 6. Indeed, you could say a competition of sorts is going on between those two powerful and opposing sets of forces which respectively wear down and build up Earth's crust, and thus create and alter the natural landforms that give character to Earth's surface. Gradational forces may not have the cataclysmic pizzazz of earthquakes and volcanoes, but their results, as shown in Figure 7-1, may be truly grand.

Figure 7-1: The Grand Canyon is a product of gradational force.

Listening to the reactions of people who are seeing the Grand Canyon in person for the first time is always extremely interesting. Some gasp. Some say, "I had no idea!" Others ask, "What's for lunch?" Okay, so not everybody is mightily impressed.

But they ought to be! The beauty is spectacular, and the scale is grand. In Grand Canyon National Park, the featured attraction is a mile deep and 10 to 18 miles across, rim to rim, depending on where you measure from. By way of explanation, *carved* is a verb you see a lot — as in "The Colorado River *carved* the Grand Canyon." Phooey. The river didn't *carve* diddly. Instead, it *carried away* every last ounce of rock and soil that once occupied the space that is now the Canyon.

Now take a look at the Appalachian Mountains, even though it may seem like a complete change of subject. The Canyon "goes down" while the Appalachians "go up," but not very grandly. Indeed, while they sport a decent peak or two, nobody is going to call them "The Grand Mountains." But they used to be. As noted in Chapter 6, geologists estimate some of the Appalachians were once 30,000 feet high. That's higher than Mount Everest, the highest mountain on Earth. But those once-mighty mountains got worn down and *carried away,* mainly by runoff from precipitation, making molehills out of mountains.

And so the Grand Canyon and the Appalachians — two very different landforms — turn out to have something fundamentally in common. A process of removal has shaped both of them. That is, both have been shaped by gradational force.

Getting Carried Away

Just how does gradational force work, as for example when it turns a 30,000 foot high mountain of yesteryear into a 3,000 foot high mountain of today? Basically, it's a two-part process. Gradational force is part of nature, which has at its disposal mechanisms (described in the next section) that can break great big rocks into tiny bits of rocks that are easily transportable. After that, another set of mechanisms picks up the tiny pieces and carries them away. These two sets of mechanisms are known respectively as *weathering* and *mass wasting.* It may take them hundreds or thousands or even millions of years to turn a mountain into a molehill, but nature is in no rush. Indeed, it has all the time in the world.

Weathering the Earth

Weathering refers to the natural processes that break rock into smaller and smaller pieces. Weathering can be broken down into two types, *mechanical* and *chemical.*

Mechanical weathering

Mechanical weathering is the disintegration of rock and other solid earth material by physical means. Here are the major mechanisms used to accomplish this:

- ✔ **Frost action.** Some rocks allow water to occupy space between particles. If the water freezes, the resulting ice crystals may exert outward pressure that cracks the rock.

- ✔ **Hot-cold fluctuation.** In some locations, rock may be exposed to extreme temperature fluctuations. Over time, alternating heat and cold causes expansion and contraction, respectively, and may result in some parts of the rock breaking off.

- ✔ **Root action.** Fine, delicate roots may find their way into cracks or spaces within rock. As the roots grow, it may exert pressure sufficient to break the rock.

- ✔ **Abrasion.** Chips may break off when a hard object scrapes or rubs against rocks. During the Ice Ages, gigantic glaciers pulverized and ground up untold tons of bedrock, converting it to pebbles, gravel, and soil particles that are still with us. This action continues with today's glaciers, but on a much-reduced scale. Similarly, scraping or rubbing occurs when one piece of rock strikes another. This may happen as a rock rolls downhill, hitting other rocks as it goes; or when high winds send sand particles smashing into other rock particles; or when a swiftly flowing river causes one rock to hit another.

Chemical weathering

Chemical weathering is the disintegration of earth material by chemical means. Here are the major mechanisms used to accomplish this:

- ✔ **Rusting away** *(oxidation)*. Lots of rock contain bits of iron. When oxygen combines with the iron, rust develops and destabilizes the rock, possibly leading to its break-up.

- ✔ **Dissolution.** Direct and prolonged contact between water and certain rock minerals may cause the latter to decompose and contribute to the ultimate break-up of rock.

- ✔ **Carbonation.** Atmospheric carbon dioxide may dissolve in rainwater and create a weak carbonic acid. Long-term, this may also contribute to dissolution of certain elements in rock and contribute to its ultimate break-up.

By itself, weathering does not create a Grand Canyon or turn a mountain into a molehill. Instead, by converting large immobile pieces of rock into small transportable ones, it makes possible the movement of surface materials in ways that create or alter landforms.

Getting down and dirty: Soil

One of the principal products of weathering is *soil,* which by definition is a collection of earth particles that are no more than 2 millimeters in diameter. Soil provides nutrients to plants, without which they simply would not thrive. The key to this is the process called *osmosis,* which is the transfer of nutrients from soil to plants through root membranes. When precipitation seeps into soil, it mixes with mineral and organic matter and becomes a kind of a "nutrient soup" that coats soil particles. Plants come into contact with soil by means of their roots, which take in the "soup" by osmosis, and thus receive nutrients from the very substance in which they grow.

A *fertile* soil is one that makes lots of nutrients available to plants. An *infertile,* or poor, soil is one that does the opposite. Soil fertility varies geographically and is one of several important elements that explains why the population of a particular country may be better nourished than people who live elsewhere. Factors that determine soil fertility include the following:

✔ *Parent material.* This is the bedrock from which soil is derived. It can be extremely hard, like granite; or fairly soft, like limestone; or in between. Generally, soft rock is best because it weathers easily and produces much more soil than harder parent material. Thus, the nature of the bedrock that underlies an area may have much to do with the fertility of the soil.

✔ *Particle size.* Soil particles can be large (sand), tiny (clay), or in between (silt). Sandy soils tend to be infertile because the "nutrient soup" tends to seep through rather than be held and made available to plants. Clay soils tend to be infertile because particles are spaced tightly together, making it difficult for the "soup" to seep in or roots to penetrate. Silty soils, in contrast, tend to be very productive because they both admit and hold good amounts of "soup," as well

as roots. The bottom line, therefore, is that moderate-size particles generally make the best soil.

✔ *Climatic conditions.* Soil fertility tends to be best where the climate is not too hot, not too cold, not too wet, or not too dry. High heat speeds up organic decay, basically wasting a high volume of nutrients before soil can make them available to plants. Low temperatures slow down decay, and again render nutrients unavailable to plants. Very wet conditions flush away nutrients in soil (a process called *leaching*), while very dry climates produce very little "soup" to surround soil particles. In contrast, climatic moderation (such as occurs in the middle latitudes) encourages nutrient accumulation and retention in soil, and thereby enhances fertility.

✔ *Profile depth.* This is the vertical distance from surface to bedrock — "thickness," in other words. Thick is preferable for the simple reason that more soil is present. As is the case with other factors, thickness varies geographically. Soft-parent material may be responsible for locally thick soils, and so, too, the process of deposition, as when a silt-laden river overflows its banks and adds a new layer of sediments to the flooded countryside (also, see "Deposition" in this chapter).

✔ *Plant cover.* Organic matter contributes greatly to the quality of topsoil. Thus, plant cover is an important determinant of the geography of soil fertility. Generally, grasslands are best because their fine roots and root hairs readily decompose and are in the soil to begin with. Leaf-fall from trees has good nutrient potential, but these are deposited on top of the ground and therefore are dependent on climatic factors (rainfall and temperature) to mix with topsoil.

On a totally different matter, weathering is fundamental to the creation of soil, which is discussed in some detail in the "Getting down and dirty: Soil" sidebar. Suffice it to say here that if you think gradation and weathering are relevant only to earth-science geeks, then guess again. It's critically important to *you;* without it, there would be no soil, and therefore next-to-no vegetation, and therefore next-to-no food.

Wasting away

JARGON ALERT

Mass wasting is the movement of particles that are products of weathering. Thus, it's the aspect of gradation that is most directly concerned with creation and alteration of landforms and has two discrete components:

✔ **Erosion:** This is the removal of *particulate matter* (pieces of soil and rock) from a particular location. Thinking back to the Grand Canyon, running water removed — or *eroded* — material from the space now occupied by the canyon. Of course, those eroded pieces didn't just disappear. Instead, they went someplace else.

✔ **Deposition:** This is the putting down (or coming to rest) of eroded materials. Returning once more to the Grand Canyon, all the eroded material that was carried away by the Colorado River was eventually deposited either down-river or settled in the Gulf of California, into which the Colorado empties.

The difference between erosion and deposition is rather like that between pick-up and delivery. Material is taken from one place and put in a different place.

Changing the Landscape

Beauty, as they say, is in the eye of the beholder. You can find it in art museums throughout the land and, as far as geography is concerned, in the land itself. Sculptures are made by removing material that once surrounded the final forms. Compositions are achieved by bringing together materials that were formerly separate. In a similar manner, erosion and deposition (see previous section) are creative processes. Like standard works of art, the results may inspire us or, frankly, leave us unmoved. Either way, the creative power of gradation produces works that we live on and, in the case of soil, cannot live without.

In the real world, nature has four means, or *agents,* at its disposal to quite literally carry out erosion and deposition: gravity transfer, flowing water, glaciers, and wind. Although a glacier is a form of flowing water, the gradational actions of solids (ice) and liquids is sufficiently different to merit separate treatment.

Staying grounded: Gravity transfer

Gravity is constantly "pulling down" on surface material. If not somehow restrained, therefore, particulate matter ranging in size from soil to boulders may move downslope in an act of *gravity transfer*. This process can be awesome, as in the case of a landslide. Much more common, however, are the decidedly unspectacular minute movements (*soil creep*) of small particles, and the occasional pebble and rock that roll a bit downhill. Given enough time, however, the cumulative effect of gravity transfer may be really noticeable. Uplands are eroded and reduced, while deposition creates new landforms, as when rock materials accumulate at the bases of mountains or cliffs to form talus cones, as shown in Figure 7-2.

Figure 7-2: Talus cones have formed at the base of the cliffs.

Going with the flow: Water

Flowing water is far and away the principal agent of erosion and deposition. The extent to which it can rearrange the landscape is largely dependent on three things that vary widely.

- ✔ **The amount of water.** This is a no-brainer. The larger a river or wave, the greater its ability to move surface materials.

- ✔ **Velocity.** The faster water travels, the greater its ability to pick up and move surface material. On land, *gradient* (the steepness of a slope) is a major determinant of speed. The steeper the inclination, the faster water travels. Similarly, storms at sea greatly increase the size and velocity of waves, and therefore greatly increase their impact on coasts.

✔ **Surface cover.** Generally, the more open or bare a landscape, the greater is the likelihood of its alteration by flowing water. Vegetation inhibits erosion by slowing down flow speed and by generating root systems that hold soil in place. Bare ground, in contrast, is rather at the mercy of water and velocity. Thus, one of the great geographical ironies is that flowing water is the principal agent of landscape change in dry areas, where bare ground is very common. Rain and streams may be scanty in arid and semi-arid regions, but it doesn't take lots of flowing waters to make major impressions on the land in those areas. Witness the Grand Canyon, for instance.

Any volume of water that interacts with earth's surface can produce mass wasting. That includes small-scale phenomena like raindrops and rivulets. Thus, run-off on exposed soil in an agricultural field or even a backyard garden may erode soil and result in *gullying* that produces very miniature versions (an inch or so deep) of the Grand Canyon. From the perspective of geography, however, it is large-scale phenomena, like rivers, waves, and ocean currents, that are of greatest interest.

Rivers and streams

The characteristics of individual rivers and streams vary greatly. In very general terms, however, one may think of a river system as originating in highlands, gathering the waters of tributaries as it snakes through foothills, and then flowing through a low-lying coastal plain as it approaches the seas (illustrated in Figure 7-3). The nature and effects of mass wasting tend to be very different in each of these settings. Here is what happens in each setting:

✔ **Highlands setting:** By their nature, highlands are high above sea level. Steep gradients are most likely to occur there, resulting in rapidly flowing streams that erode their beds and carry away weathered material with comparative ease. This process creates valleys. In highlands, V-shaped valleys are fairly commonplace, and a sure sign that erosion rather than deposition is playing the largest (in fact, almost exclusive) role in changing the landscape.

✔ **Foothills setting:** In foothills, river gradients tend to be much less steep, resulting in decreased velocity and, therefore, decreased erosion of streambeds. In fact, erosion of valley walls may exceed erosion of the riverbed. As a result, the profiles of foothills valleys may assume the shape of a somewhat flat-bottomed V — or rather _/ — with much of their floors being occupied by land rather than river.

While the fringing slopes — that is, the lines of the V — provide evidence of the continued presence and power of erosion, the appearance of a flat and comparatively (versus the highlands) expansive valley floor is testimony that deposition has been an active player in shaping the landscape. Not uncommonly, the width of these valley floors is sufficient to accommodate modest-size towns and substantial agricultural activity.

✔ **Coastal plain setting:** In the coastal plain, rivers are at their peak volume by virtue of so many tributaries having added their waters to the combined flow. Velocity is fairly slow, however, since by definition coast plains are low-lying and flat. Here the individual rivers flow through broad *floodplains* bound by diminutive valley walls. The rivers themselves are bound by even less diminutive natural *levees* that consist of sediments deposited during past episodes when the river overtopped its banks.

Indeed, "floodplains" aren't so-named for nothing. Because the landscape is so flat and expansive, flooding tends to affect a very wide area. Such events are beneficial to the extent that silt and other sediments in floodwaters are deposited and enrich the soils. On the other hand, flat and productive land attracts people and enterprise, and thus virtually guarantees that floods have a major impact on life, property, and infrastructure (see sidebar).

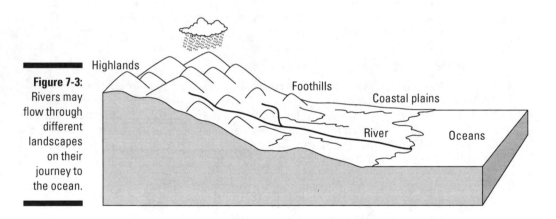

Figure 7-3: Rivers may flow through different landscapes on their journey to the ocean.

When rivers empty into the sea, they disgorge eroded sediments that have been carried along. Where coastal waters are fairly calm, these particles may be deposited at the rivers' mouths, and progressively accumulate and form a delta, as shown in Figure 7-4. This landform is so-named because many of them — most notably that of the Nile River — assume a roughly triangular shape reminiscent of the Greek letter delta (Δ).

Figure 7-4: Deltas are formed from sediments deposited at river mouths.

Time ⟶

The floodplain: Land of promise and pitfalls

Floodplain refers to flat lands beside rivers that are prone to flood when the waters overflow their banks. Their extent may vary from a few feet to many miles on either side of the watercourse. The latter is common in the case of large rivers flowing through coastal plains.

Floodplains often are characterized by rich, fertile soils (*alluvium*) that have been deposited by past floods. For thousands of years, people have been attracted to these areas due to their favorable prospects for good harvests. Indeed, the attraction is so great that several alluvial regions — including the lower Nile, Ganges, Huang, and Yangtze (Chang) Rivers — have long supported very high human-population densities. And therein lies a major predicament: The alluvial lands that attract and nourish so many, may also be the scenes of terrible flooding and uncountable drownings. For that reason, the Huang (Yellow) River, which provides irrigation for millions of acres of cropland, is also called "China's Sorrow."

In modern times, engineering has sought to control (and ideally eliminate) flooding by building high artificial levees (banks) intended to make rivers stay put. While these defenses generally work, they occasionally fail in the face of record high water, as happened spectacularly and tragically in the cases of the Mississippi and Missouri Rivers during the summer of 1993. Indeed, critics claim that, for two reasons, these bulwarks merely guarantee that the inevitable flood will be exceptionally and unnecessarily disastrous. First, they say, the human earthworks (levees) create a false sense of security that encourages construction and settlement in patently hazardous areas. Second, artificially high levees negate the natural "sponge effect" of long, wide floodplains, and therefore magnify the effects of the inevitable flood.

Low-cost federal flood insurance is an interesting variable in relation to the floodplain. It clearly and compassionately helps to relieve the pain and suffering of people who "lose everything" in times of flooding. But critics claim its promise of compensating individual losses also encourages settlement of hazardous areas, and thus may serve to increase the very suffering it seeks to relieve.

Waves and currents

Coastal areas bordering oceans and seas witness significant gradational activity. Every wave that strikes land potentially performs mechanical weathering, however minutely. Every drop of wave water that soaks land may contribute to chemical weathering. And, of course, every wave and coastal current, if strong enough, can erode sand and other coastal particles and deposit them somewhere else. As a result, coastal zones are among those parts of Earth that are most prone to change by natural means.

Eroding cliffs

Cliffs characterize many coasts, and are a sure sign of erosion. Each wave minutely helps to weather and erode the base of those landforms, undermining the entire cliff. Eventually, a portion of the cliff collapses and the shoreline *retreats,* that is, erodes inland. Figure 7-5 shows an example of a cliff formation. The rate of erosion varies from one cliff to the next depending on its composition. For example, a cliff composed of a mix of soil, gravel, and rock retreats far more rapidly than one composed of solid rock.

"Million-dollar views" may be had from cliffs, and that often results in prized pieces of property. The fact that a cliff is an erosional feature, however, virtually guarantees loss of adjacent real estate. Property owners who build houses close to drop-offs sometimes try to artificially stabilize the cliff to prevent further erosion. While such efforts usually are effective against minor storms, they stand little chance against repeated major ones.

Figure 7-5:
Cliff erosion varies from one cliff to the next.

Eroding beaches

Coastal areas that consist of sandy beaches and dunes are among the easiest landforms for nature to erode because they consist of small particles. The number of pieces of sand that make up a beach is beyond comprehension; but as far as mass wasting is concerned, particle sizes matter more than particle numbers. Take a nice sandy strand consisting of several giga-trillions of sand particles, let a severe storm pound away at it for several hours, and the result may be a greatly diminished beach.

The implications of this for property owners are severe and have been recognized for some time. A well-known parable compares a house built on rock to one built on sand. After a storm, the former remains standing while the latter has been washed away, which indicates that people thousands of years ago understood the basics of coastal erosion.

That knowledge has not, however, deterred people from building on sand. In the United States the number of people who live along the coast has soared in recent decades. Part of this is simply due to general population growth, as a result of which coastal cities have expanded up and down their respective shorelines. Probably of greater importance, however, in explaining the extent of coastal development is the American love affair with the beach, and the growing number of people who possess the financial wherewithal to purchase vacation or retirement homes by the water. Indeed, the combination of disposable income plus competitive bidding has served to make coastal real estate among the highest priced to be found in non-urban settings.

Nature, however, is not impressed by price tags. In several instances severe storms have eroded dunes from under houses, literally leaving them high and dry, such as in Figure 7-6. In response, several coastal communities have spent large sums of money to replenish beaches by dredging or pumping in offshore sand. At best, however, these are short-term cures with paltry prospects for long-term success, given the inevitability of future storms aided and abetted by rising sea levels (see Chapter 8).

Figure 7-6:
Erosion threatens to carry away the land on which these structures stand.

Coastal deposition

Of course, all the material that is eroded along beaches and cliff fronts goes somewhere. Much may be carried out to sea and come to rest on the ocean bottom. Sand, for example, may accumulate in the near-shore environment, forming shallow rises called *sand bars*. Other materials may create or add to *spits* (small points of land that extend outward into the water), or long narrow *barrier islands* that roughly parallel the coast. The latter are particularly important elements of the East Coast of the United States. New York's Fire Island, the Jersey shore, North Carolina's Outer Banks, South Carolina's Hilton Head, Cape Canaveral, Miami Beach — barrier islands are numerous and have seen significant development in recent decades. Each, however, represents a huge collection of small particles that can be eroded far more readily than they are deposited (see the Applied Geography sidebar).

The chill factor: Glaciers

A *glacier* is a large, moving mass of ice on land. It originates when more snow falls in winter than melts in summer. If this is repeated for many years, then the annual surplus of snow compacts under its own weight and forms ice.

The massive weight of glaciers is capable of grinding up even the hardest rocks into soil particles. Since glaciers move, they are a combination earth crusher-bulldozer that perform weathering, erosion, and deposition all in one. The precise nature and results of these activities depends on whether the ice in question is a mountain glacier or a continental glacier.

Mountain glaciers

As the name suggests, *mountain* (or *alpine*) *glaciers* originate in snow that falls in mountainous areas. When large ice masses form, the slopes facilitate their downhill movement, and thus their power to erode and transform the landscape. Far up-slope, the erosional power of so much ice "gnaws away" at the mountain, turning rounded tops into pointed *horns* and nondescript ridges into jagged crests called *arêtes*.

Applied Geography: Insuring against erosion

People who buy land and build on it generally expect their property to be there tomorrow. But just in case, they buy insurance. Some policies that cover coastal real estate are issued by private companies and others by branches of government. Either way, and rather often, erosion takes its toll, so policy owners end up cashing in.

If you are thinking, "I don't own coastal property, so this doesn't concern me," then I beg to differ. Every successful insurance claim drives up policy costs, so, in that sense, everybody pays to compensate people for eroded coastal property. In recent years, however, many have

begun to question the wisdom of such insurance. Key to this is growing public awareness of basic physical geography, particularly erosion as it relates to beaches. In some states and regions, the results have been moratoriums (or talk of them) on beachfront housing, or on insurance policies that protect them. In so doing, the general public and their elected representatives have been applying a basic concept that geographers have appreciated for some time, namely, that the works of humans ultimately stand little chance of permanency in that most dynamic of natural environments, the coast.

Main valley floors are widened and deepened, changing from V-shaped valleys into larger U-shaped *glacial troughs.* Higher up, side valleys are also carved out. Less ice flows through them than the main valley, so less downcutting occurs. Once the glaciers recede, the former side valleys are left "hanging above" the main valley, and are thus referred to as *hanging valleys.* In the mountainous coastlands of Norway, New Zealand, Chile, Western Canada, and Alaska, glacial troughs have filled with seawater, resulting in steep-sided ocean inlets called *fiords.*

Glaciated mountains have a spectacularly rugged look about them. Perhaps not surprisingly, therefore, such areas tend to be major magnets of travel and tourism. Also, one must remember that these landforms were created by erosion, which leads one to ask where all the removed material went. The answer is that it was deposited downslope, forming or mixing in with soil or, in the case of fiords, ocean bottom.

Continental glaciers

Continental glaciers build up over large land masses in general, as opposed to mountains in particular. At the height of the last Ice Age, some 20,000 years ago, they covered substantial portions of North America, as well as fair portions of Eurasia. The present ice caps of Greenland and Antarctica (see Chapter 8), which are more than 2-miles thick in some places, offer insight into continental glacier scale and movement.

In North America, massive ice sheets (as seen in Figure 7-7) slowly built up over many years in what is now Canada. Eventually, the unimaginable weight and volume caused the ice to "ooze" outward in response to the pressure, at which point, the ice technically became a glacier. As the ice sheet advanced, its immense weight weathered the underlying Earth and *gorged out* (eroded) low-lying landforms through which it passed. Thus were formed, for example, the beds of the Great Lakes, the Finger Lakes (New York State), and numerous other future water bodies. Much of Canada was scoured by that process, which explains the relative lack of soil today over parts of that country, as well as the presence of thousands of lakes.

Eventually, of course, all of the eroded material being carried or pushed along by the ice got deposited, particularly as climates warmed and the ice sheets waned. In some places, these deposits of rock, sand, and other debris (called *glacial till*) merely coated the surface. In others, however, major accumulations occurred, resulting in landforms called *moraines.* Long Island and Cape Cod are noteworthy examples. These were created by the bulldozing effect of the ice sheets and the transporting and deposition of debris beneath them.

No Glaciers

Ice Age Glaciation

Southern Limit of Glaciation

Figure 7-7:
Extent of
continental
glaciers in
North
America
during the
last Ice Age.

Generally, as the ice receded across the United States, it left behind a blanket of stones, gravel, and soil. Melt water issued from the retreating glaciers in innumerable streams that carried and deposited fine soil particles, which further transformed the post-glacial landscape. Today, thousands of years later, much of those materials underlie America's agricultural heartland and are important factors in explaining its productivity.

Making a deposit: Wind

Given sufficient velocity, wind can pick up soil particles and carry them long distances before they are deposited. This was dramatically demonstrated in the 1930s when portions of the American High Plains endured the Dust Bowl phenomenon. Wind-related erosion and deposition occurs most commonly in arid and semi-arid areas where ample bare ground is exposed to the atmosphere. In such venues, dust and sand "storms" of varying intensity are not unusual. *Sand dunes* are a classic landform that results from the inevitable deposition.

In a few non-arid parts of the world, substantial (that is, meters deep) deposits of wind-blown silt, called *loess,* are present. Because these are very fertile soils, loessal areas are among the most agriculturally productive to be found anywhere. Large areas of the American Midwest are covered by loess, as illustrated by Figure 7-8. Some of it probably blew in from arid areas. The common belief is that most loess, wherever they are found, originated long ago in dry glacial till that was exposed to wind after the ice sheet retreated.

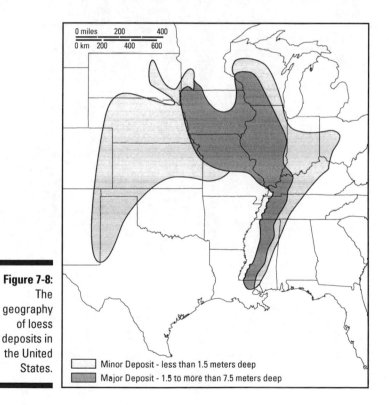

Figure 7-8:
The geography of loess deposits in the United States.

0 miles 200 400
0 km 200 400 600

Minor Deposit - less than 1.5 meters deep
Major Deposit - 1.5 to more than 7.5 meters deep

Chapter 8

Water, Water Everywhere

*T*he sound of the Earth is *ker-ploosh*. Nearly 70 percent of our planet's surface is covered by water, and the vast majority of that water is ocean. In addition, water is in the air you breathe and mixed in with the soil underfoot. Then you have lakes and rivers and icecaps. So, it really isn't a stretch to say, "Water, water everywhere."

The geography of water is vital to our lives. It determines where people live, the shape of our world, the content of the atmosphere, who can fish where, where travel is permitted, and how to supply the population with enough to drink. You may never look at a glass of water in the same way!

Many geography books don't include a chapter on water. Instead, the authors opt to sprinkle the subject throughout their books. Obviously, I've decided to go against the flow and put it all here in one big puddle, with one exception: I'm going to hold off discussing ocean currents, which have a major influence on climate, until the next chapter.

Taking the Plunge: Global Water Supply

Although water is everywhere, it's very unevenly distributed. The oceans account for 97.25 percent of the global water supply. Ice caps account for another 2.05 percent, and together that makes 99.3 percent. Of the 0.7 percent that remains, 0.68 percent is groundwater. Most of that is mixed in with soil, but a small portion is potentially available to humans as well water. And that leaves 0.02 percent, which comprises all the world's lakes, all the world's rivers, and all the vapor in the atmosphere.

Where did all that water come from?

Actually, *all that water* isn't all that much considering Earth as a whole. Water may cover 70 percent of Earth's surface, but it accounts for only about 0.5 percent of the planet's weight. Oceans, on average, are about 2 to 3 miles deep, but the average distance from Earth's surface to its center is 3,960 miles. So the deepest of the deep blue seas is but a shallow veneer of surface material.

Apparently, all that water was here from the beginning. When Earth was a newborn fireball, its water was mixed together with other planetary matter. Indeed, probably much, much more water was on the inside of the planet than on the outside. Any substance that can exist as a solid, liquid, or gas is rather amazing. That's water. Because early-Earth was so hot everywhere, water was then in a gaseous state instead of liquid. Gases are lighter than solid

matter, so they want to rise. In time, the gaseous water inside the Earth migrated (rose) to the outside and into the primordial atmosphere. This migration continued until the crust cooled (blocking the migration of additional internal water vapor to the surface), although subsequent volcanic activity may be thought of as planetary "burps" that spew additional water into the atmosphere, along with a lot of other stuff.

Eventually the water vapor in the atmosphere condensed and formed rain. Some scientists believe that downpours began very early on. Others believe that didn't happen until eons later, after the crust cooled. But most agree that the vast majority of Earth's surface waters originated in rain that fell from the sky over a very, very long period of time.

What these data tell us is that the vast majority of the Earth's water is unavailable for human use. In some countries, there are "desalinization" plants that — as the name implies — remove the salt from seawater and produce potable fresh water. The process is rather expensive, however, and helps satisfy the needs of only a few localities — coastal cities, mainly. As a meaningful solution to the global water crunch, it's simply not feasible given the current level of technology. And if the amount of potable water isn't worrisome enough, the geography of supply is often way out of whack with the geography of need. Fresh water is often abundant in areas where human need for it is scarce. And fresh water is often scarce where human need is abundant. The following sections show you just how the global water supply is broken down.

Those ice caps are really cool!

Earth's "ice caps" are actually continental glaciers, whose origins are described in Chapter 7. Cool? They sure are. Ice caps account for about 2 percent of Earth's water. That may not seem like much, but as a portion of all the water that exists on this planet, 2 percent turns out to be a lot of wet stuff — or actually hard stuff, because it's water in a solid state.

You want data? Antarctica's ice cap is as deep as 15,760 feet. If you stood at that spot, you would not literally be *on top of* Antarctica. Instead, you would be on top of a 3-mile-thick piece of ice that is on Antarctica (see Figure 8-1). Indeed, the total volume of that ice cap is estimated to be about 30 million cubic kilometers. That's a lot of margaritas.

Figure 8-1:
A portion of Antarctica's ice cap.

Want more? Well, there's Greenland, which is one of the great misnomers in the history of real estate. It really ought to be named Whiteland, because close to 99 percent of it is covered by ice cap. Actually, Greenland's ice cap is only about 10,000 feet deep at its thickest, so it's practically minor league compared to Antarctica.

Getting out: Oceans, seas, gulfs, and bays

Most of earth's water is oceans, including seas, gulfs, and bays. The difference between these terms is basically a matter of size and location. *Ocean* comes from the *Okeanos* of Greek mythology, which was a river thought to encircle the Earth. And indeed, *the ocean* is a continuous body of water that encircles the land, but it also consists of a handful of divisions that are also referred to as oceans — Atlantic Ocean, Pacific Ocean, Indian Ocean, and Arctic Ocean. So *ocean* is the big enchilada.

Seas, gulfs, and *bays* are parts of the ocean that adjoin land bodies. Generally, seas are ranked second (after oceans) with respect to size. They may be relatively enclosed by land, as is the Mediterranean Sea, or they may be "open" to the ocean, as is the Arabian Sea. A *gulf* is a part of an ocean or sea that extends into the land. Dictionaries suggest that gulfs are smaller than seas but bigger than bays. A bay may be defined as an inlet of a sea or gulf. For example, Tampa Bay is an inlet of the Gulf of Mexico.

Why are the oceans salty?

Dissolved mineral salt is the key. As water flows to the seas, it comes into contact with lots of rock and rock particles. These contain mineral salts, minute quantities of which are dissolved and carried along by running water. It's probably in the stuff you drink, but because the salt content is so low, you don't notice anything peculiar.

Ultimately, that water with its low salt content joins the sea. The sun continually evaporates sea water, producing vapor that becomes future rainfall. But here's the punchline: It's the ocean water that gets evaporated and not its salt content. So during evaporation, the salt gets left behind, only to have "new salt" added to it as freshwater eternally runs to the sea. Give this process a couple million years (which it did), and the result is an ocean of water too salty for human consumption.

How about de-salting the waters? That certainly is possible. *Desalination* (distilling sea water to remove the salt) is a fairly simple process, and goodness knows the oceans contain more water than humans will ever need. But de-salting seawater in the copious quantities that cities require is very, very expensive. Few can afford it, so desalination is not a particularly viable option.

That's by the book, of course. In reality, things aren't so neat and tidy. Considering the Indian Ocean and parts thereof (see Figure 8-2), for example, you can see that the Bay of Bengal is bigger than the Andaman Sea and the Gulf of Oman. When it comes to these place-names, it's kind of like being told "size matters" and then finding out it doesn't. Basically, what happened is that, way back when, some explorer or mapmaker simply labeled a water body and the name stuck. Maybe that person didn't appreciate the true size of the feature being named or didn't appreciate the nuances of vocabulary. Whatever the case, the result is a dictionary that suggests a definite rank order with respect to size of seas, gulfs, and bays, and a world map that says it just ain't so.

Figure 8-2:
The Indian
Ocean
and parts
thereof.

Coming inland: Lakes

A similar brand of confusion reigns with respect to inland bodies of water. A *lake* is a body of water completely surrounded by land. But when you open an atlas and browse the lakes, you find some are called seas. For example, the Caspian Sea and Aral Sea are in Asia. The Salton Sea is in California, and the Dead Sea is between Israel and Jordan. Logic would suggest that a "sea" should be bigger than a "lake." In reality, every one of the Great Lakes is larger than the Dead Sea, and so are a bunch of not-so-great lakes.

The explanation is that the previously-named seas (which are really lakes) are salty. Each occupies a *basin* — a depression in the landscape. Fresh water flows in, but nothing flows out. So you end up having an ocean in miniature. That is, the rivers bring minute quantities of dissolved salts to the lake/sea, which is essentially a dead-end repository. When the sun evaporates the surface waters, the salt gets left behind. Keep this up for many, many years, and what you have is a lake whose waters are not only salty, but are even saltier than your average ocean water. That's the way it is with the Great Salt Lake, which apparently has every right to be called a sea but is not. Again, somebody way back when gave a name to a water body, and the label stuck regardless of the logic.

Shaping Our World: Oceans

Being the largest supplier of water on the planet, oceans have a big responsibility. This is where major transportation takes place between countries, where wars are sometimes fought, and where food is caught, not to mention the physical effects that oceans have on the land they wash up upon.

Going where the action is: The continental shelves

Off the coasts of continents are relatively flat expanses of ocean bottom that average about 300 feet in depth — shallow by oceanic standards. Such a feature is called a continental shelf (see Figure 8-3). The distances that the shelves extend from the shore vary. While the continental shelf off California extends 2 to 3 miles, the one off Newfoundland extends seaward some 200 miles.

Altogether, the continental shelves define what is arguably the most economically important part of the ocean. A growing human population translates into a need for more food (especially stuff that is high in protein) and more mineral resources. Increasingly, oceans are being looked to as sources of supply. And in that regard, the continental shelf is where the action is.

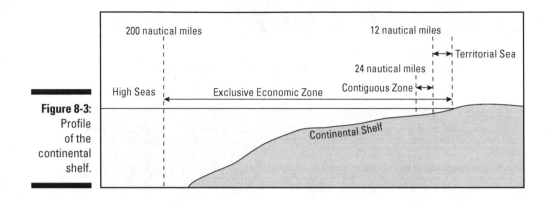

Figure 8-3:
Profile
of the
continental
shelf.

Something very fishy going on

If you were a fish and lived in the ocean, then chances are you would hang out over the continental shelf, just like almost every other fish. Schools are in session there for good reason — lots of food is available. Here's how the food gets there:

- ✔ **Plant life:** Because shelf water is shallow (compared to mid-ocean depths), some sunshine hits bottom and gives rise to plant life that serves as food for small fish that serve as food for bigger fish, and so forth.

- ✔ **River flow:** Rivers empty onto the continental shelves. Their flow typically contains a lot of organic matter (especially dead, decaying, or dissolved plant parts), which adds to the abundance of fish food.

- ✔ **Vertical mixing:** Wave action and turbulence produce a considerable amount of vertical mixing over the continental shelves. As a result, organic matter gets distributed over the depths, making for a very robust feeding environment.

Fuel for thought

Petroleum and natural gas underlie some areas of the ocean bottom, just as they are located under some areas of dry land. In the oceanic setting, these resources are found exclusively under continental shelves because they alone in the marine environment are composed of the geological features in which oil and gas are found. So when you see photos of offshore oil rigs, you are looking at an important economic activity that coincides with a continental shelf. Again, that's where the action is.

Claiming ocean ownership

The existence of all the marine goodies leads to the rather important issue of ocean ownership. That is, who owns them? How far offshore does a nation's sovereign territory extend — if at all? For countries that have coastlines — and for ships at sea — having an answer is important for the following reasons.

- **National security:** Countries have the right to defend themselves from attack or intrusion. It's crucial to be able to determine, therefore, the point at which a ship "crosses the line."

- **Police power:** Criminal activities can take place on water as well as land. Drug smuggling is a key example. Legally, however, the police and Coast Guard can only board vessels at sea within their jurisdictions. Thus, it is of some importance to clearly define how far out to sea those jurisdictions extend.

- **Trade and commerce:** Thousands of freighters and tankers ply the ocean. The captains and navigators who set ships' courses need to know where they can "sail as they please" and where they need permission by virtue of having entered a foreign country's territorial waters. Only by defining the extent of ocean ownership can that be determined.

- **Resource ownership:** Disagreement between countries concerning ownership of or access to resources can lead to conflict. "Marine goodies" mentioned earlier are potential sources of contention. Therefore, clear definition of the extent of ocean ownership may prevent conflict over oceanic resources.

- **Dire straits:** The oceans contain several *straits* — narrow waterways that separate land bodies. About a dozen of them are major bottlenecks as far as international shipping is concerned. It is important to determine whether these are international waterways through which ships of all nations may freely pass, or if they belong to the countries that border them — and which would therefore have the right to deny passage or charge tolls.

So a tragedy occurs. And indeed it is a double tragedy, because it didn't have to happen. If one or more coastal countries had had jurisdiction over the fishery — the commons — then, ideally, they would have enacted and enforced resource management rules designed to keep fish populations and fish harvests in balance. And in the end, all concerned would have benefited — the fishermen, the consumers, and most importantly, the fish.

For decades the United Nations (and the League of Nations before it) sponsored conferences on "The Law of the Sea" to determine the nature and extent of offshore jurisdiction as well as other issues related to ocean use. In 1982, those efforts resulted in a draft treaty entitled the *United Nations Convention on the Law of the Sea* (UNCLOS), which became international law in 1994. The jurisdictional provisions are depicted in Figure 8-3. If you think of yourself as the coastal country in that diagram, then the following list details what the different zones mean to you.

- **Territorial Sea:** Adjoining your coastline is a zone known as territorial sea that extends 12 nautical miles (nm). (A *nautical mile* equals 1.15 "regular" miles.) This area is all yours to do with as you please. Vessels that might engage in fishing or mineral extraction in this area are subject to your licensing and permission. Vessels that just want to pass through have the *right of innocent passage,* meaning they are free to travel provided they don't threaten your security or laws. Warships of other countries have the right to be here, but in normal times that is considered unnecessarily provocative.

- **Contiguous Zone:** The Contiguous Zone extends another 12 nautical miles past the border of the territorial sea. You don't own this area in all respects, but you can enforce your customs, immigration, and sanitation laws within this area. *Hot pursuit,* chasing and boarding vessels suspected of involvement in illegal acts, is also allowed.

- **Exclusive Economic Zone:** Exclusive Economic Zone (EEZ) extends as far as 200 nautical miles from the coast. All fishing and mineral exploration in this area is under your control. The presumption, of course, is that you exercise this power wisely for the management of marine resources. Also, it's no accident that the dimension of EEZ places almost all of the world's continental shelf under somebody's official jurisdiction.

- **The High Seas (International Waters):** *The High Seas* (also known as *International Waters*) extends beyond the EEZ. They are open to vessels of all nations for whatever purpose. Thus, everybody enjoys the right of innocent passage in this area. In theory, deep ocean mineral resources are for the benefit of all the peoples of the world. In reality, the minerals down there are at the disposal of a handful of countries that possess the technological wherewithal to go get them.

The various offshore zones may appear straightforward, but in the real world they often don't work. Consider the United States and Cuba, two countries that aren't exactly bosom buddies. Each country is theoretically entitled to what is shown on Figure 8-3. But Key West, Florida, is only about 90 miles from Cuba. Adjustments had to be made, therefore, beginning with a zone of High Seas in between to guarantee the right of innocent passage to ships of all countries. After that, the U.S. and Cuba were each allotted equal, but reduced, amounts of the other zones shown on the diagram.

When you give the world map a major look-see, you find virtually dozens upon dozens of other watery expanses where the diagram "doesn't work." Each was reconciled on a case-by-base basis. Straits were given special attention to guarantee the right of innocent passage.

Getting a rise out of oceans

Sea levels are rising. The ice caps on Greenland and Antarctica, plus a majority of the world's glaciers, are slowly shrinking. The shrinking is a by-product of *global warming* (also called the *greenhouse effect*), which you can read about in Chapter 18. As the glaciers slowly recede, their melt water returns to the oceans, which rise. Perhaps the most important word there is "slowly." Unless you live very close to sea level, you have no cause to have a nightmare about the ocean rising around your ankles. I would caution you, though, to think very carefully about beachfront real estate as a long-term investment.

How high the oceans will rise is open to debate because nobody knows for certain the future course and severity of global warming. But consider the past as a portent of possibilities. About 18,000 years ago, at the peak of the last Ice Age, sea levels were about 350 feet lower than they are today — because so much water was "locked up" on land in the great glaciers. About 400 years ago, the English founded Jamestown, Virginia, and built a wharf. Today, that wharf is about 12 feet under water, a victim of rising sea level.

Geographically, this process is very important. Perhaps the most basic element of geography is the familiar outline of the continents and other major land bodies that you see on a world map. For the last 18,000 or so years, the world map has been changing, with the ocean on the up-and-up. And this trend will continue for the foreseeable future. Thus, you may legitimately think of today's world map as a single frame of a very long movie. You know the plot: Oceans are rising. As for the ending . . . well, several outcomes are possible. All involve higher sea levels. Even if the change is slight, the impact will be significant. Even if the rise is just a couple of feet in the next century, millions of people will feel the effects. The highest elevations of several island republics (The Maldives in the Indian Ocean, for example) are "just a couple of feet" above sea level. Much of the Bahamas is "just a couple of feet" above sea level, and so too are large areas of Bangladesh — one of the most densely populated countries in the world. And literally dozens of populous port cities around the world have millions of people and trillions of dollars of infrastructure "just a couple of feet" above sea level. Significant change is on way. The map will look different. The world's geography will be different.

Getting Fresh with Water

Water, water everywhere. While the oceans may be the biggest water bodies out there, they're not necessarily the most important — it's time to talk about some drops you can actually drink. And that entails the *water cycle* (or *hydrological cycle*), which, as far as physical processes are concerned, is a doozy. See Figure 8-4 for a look at the water cycle. You simply cannot live without it, and neither could any other living thing that requires fresh water because the water cycle is the world's one and only producer of fresh water — no water cycle, no fresh water — it's as simple as that.

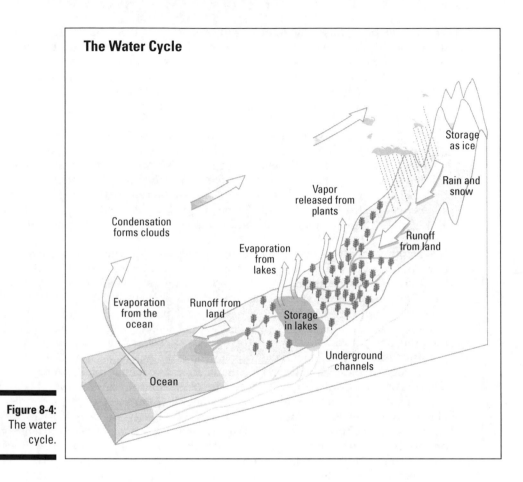

The Water Cycle

Storage as ice

Rain and snow

Vapor released from plants

Condensation forms clouds

Runoff from land

Evaporation from lakes

Evaporation from the ocean

Runoff from land

Storage in lakes

Underground channels

Ocean

Figure 8-4:
The water cycle.

In addition to the stages described in the next section, the water cycle has two overriding characteristics that are good to keep in mind. Here they are:

- ✔ **It really is a cycle.** The water really does go round and round just like the diagram suggests. "Cycle," therefore, very appropriately describes what's going on.

- ✔ **The cycle is a closed system.** By closed system, I mean that nothing gets added to it or subtracted from it. For all intents and purposes, the amount of water on Earth is fixed — closed to change. Thus, if a lake or reservoir dries up, the water is not really "gone." Instead, it has relocated elsewhere within the system.

The stages of the water cycle

Here are the principal components of the water cycle:

- **Solar energy:** Technically, sunshine isn't part of the water cycle for the simple reason that sunlight isn't water. But it's the sun that sets the cycle in motion. The sun is the pump that gets things moving.

- **Evapotranspiration:** Whew! That's a mouthful. And it's a combination of two words: evaporation and transpiration. With regards to the first word, some water evaporates when it receives solar energy. That is, it changes from a liquid state to a gaseous state: vapor. With salty seawater, only the water evaporates, and not the salt. So salty seawater becomes freshwater vapor in the atmosphere.

  With regards to the second word, you perspire, plants transpire. Plants sweat when the sun heats them up — only it's called transpiration. Plant sweat is a key input into the vapor in the atmosphere, especially in lush, tropical areas.

- **Condensation:** Water vapor is so small that it's invisible. But when lots of individual bits of vapor cluster and combine, they become visible miniature droplets of water. This is condensation, and if it happens by the gazillions, you get a cloud. As to exactly what happens to produce condensation, well, I'm holding that until Chapter 9.

- **Precipitation:** If condensation continues, then the droplets continue to get bigger and put on weight until they are too heavy to remain suspended in the atmosphere. They then fall to Earth as precipitation. This phenomenon may take various forms, including snow, sleet, hail, and most commonly, of course, rain.

- **Run-off**

- **Infiltration**

No, I didn't forget to write something. Instead, each deserves a separate heading because, no matter where you live, one or both is responsible for your water supply.

Run-off: Going with the flow

Some of the water that falls to earth collects on the surface and begins a down-slope journey to the sea — there to complete the water cycle. Trickles join to form babbling brooks (why do brooks babble, anyway?), which join to form rivers. Basically, it's fresh water on the move, and every last drop of it is potentially available to people.

Rivers have long served as water supply systems for cities and towns. But rivers that bring drinkable water can also take away sewage and waste. That's great unless you happen to live downstream — there to discover your drinking water is no longer drinkable thanks to your upstream neighbors. And most people live down stream from somebody else, so problems afloat.

To solve the problem, many towns and municipalities have gotten into the business of capturing drinkable run-off by building dams that create reservoirs that store water that can be transferred to where people need it. These public works tend to get located in not-yet-contaminated headwaters. Sometimes, however, these headwaters are far from the people who will consume them, and therefore require construction of aqueducts to resolve the geographic difference between supply and demand. Some of the water that supplies New York City, for example, comes from a reservoir more than a hundred miles away.

Infiltration: Out of sight, not out of mind

Some of the water that falls to earth *infiltrates* the soil. That is, it seeps into the ground. Because nature has been at this for a long, long time, some lands are covered by substantial *aquifers* — subterranean accumulations of water. Better to think of these not as underground lakes but as areas of super-saturated soil or porous rock (yes, rock!) with a high water content. If your water supply comes from a well, then you live on water that has infiltrated. Well water can be suitable to drink as is, or it may require minimal treatment. But infiltrated water has two potential problems — contamination and depletion.

Contamination

Along with rainwater, chemical fertilizers and industrial wastes can seep into the soil and contaminate aquifers. One of the more horrific potential results is a *cancer cluster* — an area whose residents have a disproportionately high incidence of some kind of cancer. These areas may occur because a factory dumped carcinogenic wastes that seeped into an aquifer and eventually mixed with people's water supply.

Depletion

Quite often, people consume water from aquifers much faster than nature replaces it. Nature was putting water in aquifers long before people came along. But the amount of water in the bank can diminish quickly when people start drawing from that watery account in quantity.

Consider, for example, the Ogallala Aquifer, which underlies a considerable portion of the American High Plains (see Figure 8-5). If you reside in the U.S., then that aquifer is vitally important to you even if you live nowhere near it because a host of crops that feed people and fatten livestock are grown in great quantities in the High Plains with the use of irrigation water from the Ogallala Aquifer. And the water within it is being consumed much faster than is being replenished. Wells must be dug deeper and deeper. Theoretically, one day, they may all dry up.

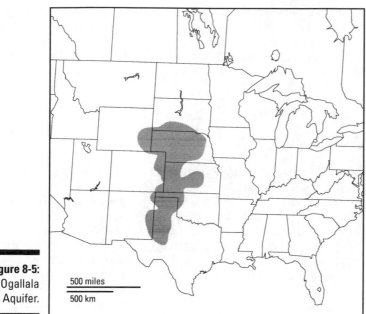

Figure 8-5:
The Ogallala
Aquifer.

500 miles

500 km

Applied Geography: Drip irrigation

In arid and semi-arid lands, water for irrigation has long been applied to fields by means of open-air ditches between rows of crops. Although this method has helped satisfy the food needs of unknown numbers of people over the millennia, it has three significant drawbacks. First, a substantial volume of water may evaporate before it reaches the plants. Second, the amount of water that seeps into the soil is usually far more than the plants actually need. Third, the mineral salts that build up in the soil as a consequence of evaporation may ultimately undermine the usefulness of the farmland and lead to its abandonment.

In some countries, however, these environmental effects have been largely nullified by introduction of drip irrigation. In this technique, water reaches the fields and is distributed up and down crop rows by means of thin plastic tubing that is perforated by tiny holes every few inches. Water is forced through the tubing at very low pressure; so instead of squirting out of the holes like so many tiny fountains, the water slowly drips out. Because of the tubing, very little water is lost to evaporation. Moreover, drop by drop application results in little water wasted and minimal salt accumulation. All in all, therefore, drip irrigation is proving to be an effective and fairly low cost means of making maximum use of a scarce arid-land resource.

Doom and gloom? Not necessarily. Advances in agriculture are making it possible to produce good harvests with less water. Greater conservation also is possible, as is greater use of regional rivers for irrigation. And if worse comes to worse, well, portions of the High Plains could revert to something akin to the lush natural grazing lands (the prairies) that were done away with to make way for farms.

Good to the very last drop

The bottom line with respect to run-off and infiltration is that you can't take more water out of the system than nature puts into it. Reservoirs and aquifers *can* dry up. In that event, water will still be everywhere, albeit in forms that are not readily accessible (such as vapor in the air or veneer surrounding soil particles) or in amounts that satisfy local needs. Humans number six billion and counting. More people mean more direct consumption, more irrigation, and more industrial use. How these needs will be met remains to be seen, but they will clearly reflect the geography of a precious resource.

Chapter 9

Warming Up and Chilling Out: Why Climates Happen

*H*ave you ever felt people were staring at you in a discreet sort of way? I certainly have. And I can give an example that relates directly to this chapter.

I was in a Swiss village on a sunny and warm July afternoon. The temperature was in the mid-eighties, which is why I was wearing shorts and a short-sleeved shirt. I got on a cable car to go way up into the mountains to take in the view and go for a hike. There were about 20 other people on board, and the higher we went, the more I felt their eyes. So I looked back, and in so doing quickly understood their stares.

Nobody else was dressed like me. Not even close. They had on heavy pants and heavy shirts, with sweaters and jackets at the ready. And for good reason. When we got to the top and the doors opened, a blast of cool air rushed in. I'm guessing the temperature was in the high forties. My first reaction was to feel totally stupid. Like, duh, shouldn't a geographer have been prepared for this? My second reaction was to cancel the hike and keep warm until the cable car made the return trip.

That brief journey began and ended on the side of the same mountain under the same sun. But something happened to the air temperature that made all the difference in the world. The same is true globally. You can travel the world and experience all kinds of climates. Together they constitute a vast array of atmospheric characteristics that concern temperature, precipitation, and seasonal change. But no climate "just happens." While all are products, directly or indirectly, of the same sun, factors are afoot that give each of them characteristics that really do make all the difference in the world.

The factors that determine climate in different parts of the world are of central interest to geography and are the subjects of this chapter. The characteristics, locations, and consequences of climate occupy the next chapter.

Getting a Grip on Climate

Like climate, weather is concerned with atmospheric conditions. The difference between the two is a matter of time. *Weather* refers to day-to-day conditions and changes in Earth's atmosphere. *Climate* refers to the average of weather conditions at a location over a long period of time — 30 years, as far U.S. government climatologists are concerned. Climate, therefore, is the more appropriate topic for this book because it concerns general characteristics of a location or region. Accordingly, among the things you will not read about here are tornadoes, hurricanes, thunder, lightning, hail, and other forms of short-term atmospheric mayhem that fall within the purview of weather. (For more information on weather, see *Weather For Dummies* [Hungry Minds, Inc.].)

So what factors cause the different kinds of climates to occur? (Drum roll, please.) Six determinants, which may act singly, in combination, or in opposition to each other, make climate occur:

- ✔ The angle at which solar energy strikes the Earth
- ✔ Tilt of the Earth on its axis
- ✔ Altitude with respect to sea level
- ✔ Solar absorption properties of land and water
- ✔ Ocean currents
- ✔ High and low atmospheric pressure belts

The following sections show how each bullet point creates climate.

Playing the Angles

Parts of Earth receive different amounts of *solar energy* — heat from the sun. The dosages are greatest in the equatorial realms and progressively diminish as one approaches the Poles. For this reason, rather warm climates generally dominate the low latitudes and give way to cooler and cooler climates in the mid-latitudes and polar regions.

These differences are due to the angles at which solar energy strikes the Earth at different latitudes. Figure 9-1 shows three "bundles" of sunshine whose widths are the same, so it can be assumed they contain equal amounts of solar energy. But if you examine the amount of Earth that each impacts, a key difference appears. Bundle A illuminates a much smaller area than Bundle B, which in turn illuminates a much smaller area than Bundle C.

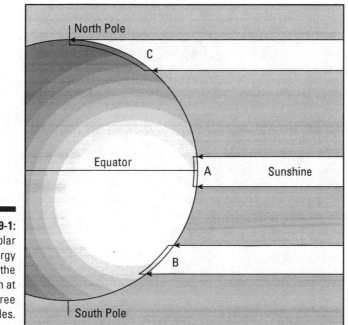

Figure 9-1:
Solar energy strikes the Earth at three latitudes.

Making hot and cold

The differences among the Bundles in Figure 9-1 are determined by the curvature of the Earth. Bundle A contains *vertical rays,* which strike Earth perpendicularly. Due to curvature, however, Bundle B strikes Earth at a sharper angle. As a result, its solar energy is spread over a larger area than is Bundle A's. Sharper still, thanks to curvature, is the angle at which Bundle C strikes Earth. And as a result, its heat is spread over the largest area of the three examples.

Absent other factors that affect temperature, Area A has the warmest climate because it has the greatest concentration of solar energy. That is, the heat in Bundle A is brought to bear on a relatively small area. In contrast, Area C has the coldest climate because the heat it receives is spread over the largest of the three areas. Intermediate conditions are present in Area B.

Let's look at some real locales and compare the climates of Manaus, Brazil, and Churchill, Manitoba (Canada). Manaus is located at about Latitude 3° South and therefore exemplifies Area A on the diagram. Churchill is located at Latitude 58° North and therefore has the characteristics of Area C.

In Manaus, the annual average temperature is 79° F. (That means that if you recorded the temperature every hour of every day for a year, the average would be 79° F). In Churchill, it's 19° F. The difference is 60° F. Manaus is much warmer.

Beware the reason! Is it because Manaus is closer to the equator? Nope. Proximity to the equator *per se* is not the explanation. Instead, the answer lies in the angle. The sun is more directly overhead at Manaus than at Churchill throughout the year. Thus, Manaus experiences a greater concentration of solar energy throughout the year and is therefore a warmer climate.

Making rain and snow

Climate is about precipitation as well as temperature. Regarding wet stuff, Manaus receives 82 inches of precipitation on average each year. Churchill, in contrast, receives 15 inches. Thus, the precipitation difference in inches is even greater than the temperature difference in degrees Fahrenheit. Again, sun angle plays a major role in these particular cases.

In Chapter 8, I discuss how solar energy causes water to evaporate and plants to *transpire* (sweat), producing atmospheric vapor, the building blocks of raindrops and of other forms of precipitation. Generally, the greater the solar energy, the greater the amount of evaporation and transpiration, which result in vapor in the atmosphere. And the greater the amount of atmospheric vapor, the greater the likelihood of precipitation. Given that Manaus experiences much more intense concentrations of solar energy than does Churchill, its atmosphere is much more humid and therefore has a higher rainfall potential. Indeed, the very heat that produces that vapor may also create an atmospheric upwelling, or *convection current,* which carries vapor to a high altitude where it cools, condenses, and falls as rain (see Figure 9-9 later in the chapter).

Tilt-a-World: The Reasons for the Seasons

Earth's axis is not, in reality, "straight up and down" as indicated by Figure 9-1. Instead, it's tilted by 23½° from the perpendicular. As a result, the vertical rays do not "stay put" at the equator as Earth orbits the sun. Instead, they "migrate" north and south of the equator at different times of year, bringing with them patterns of seasonal change (as illustrated by Figure 9-2). Take, for instance, the United States. During summer, the vertical rays move into the Northern Hemisphere, increasing its "dosage" of solar energy and producing the warm temperatures that are associated with that time of year. During winter, however, the vertical rays move into the Southern Hemisphere, greatly increasing that region's solar dosage but at the same time greatly decreasing the dosage that hits the United States.

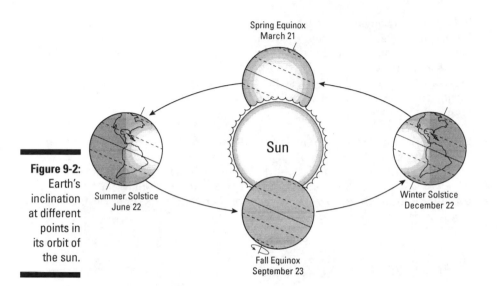

Figure 9-2:
Earth's
inclination
at different
points in
its orbit of
the sun.

Spring Equinox
March 21

Sun

Summer Solstice
June 22

Winter Solstice
December 22

Fall Equinox
September 23

Special lines of latitude

The inclination of Earth on its axis accounts for four special lines of latitude that appear on most world maps and globes (see Figure 9-3). Two of them are the *Tropic of Cancer* (23½° N) and the *Tropic of Capricorn* (23½° S). The area between them may properly be called *The Tropics*. The other two lines are the *Arctic Circle* (66½° N), and the *Antarctic Circle* (66½° S). The area north of the Arctic Circle may properly be called *The Arctic* or *The Northern Polar Region*. Similarly, the area south of the Antarctic Circle may properly be called *The Antarctic* or *The Southern Polar Region*. But why are these lines there in the first place? For now we will concern ourselves with the Tropic lines because they have much to do with defining seasonal change.

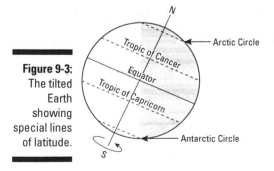

Figure 9-3:
The tilted
Earth
showing
special lines
of latitude.

N

Arctic Circle

Tropic of Cancer

Equator

Tropic of Capricorn

Antarctic Circle

S

Because of the angle of Earth's tilt, during summer in the Northern Hemisphere, the vertical rays move north of the equator as far as Latitude 23½° North. Conversely, during winter, the vertical rays migrate south of the equator as far as Latitude 23½° South. Thus, the Tropic of Cancer marks the most northerly latitude that is struck by the sun's vertical rays at some point during the year. Conversely, the Tropic of Capricorn marks the most southerly occurrence.

Parenthetically, "Tropic" comes from the Greek *tropos,* to turn. The ancient Greeks observed that during their summer the sun's vertical rays moved northerly until they reached Latitude 23½° North, whence they "turned" back to the south. Cancer and Capricorn refer to stellar constellations that were prominent in the Greek sky when the vertical rays struck one or the other tropic lines.

Defining the seasons

Four days each year, vertical rays strike either the equator or one of the Tropic lines. On two of those dates, called *equinoxes,* the vertical rays strike the equator. On the other two dates, called *solstices,* the vertical rays strike one of the Tropics. The significance of these dates is that they mark the beginnings of the four seasons of the year. The following sections show the annual cycle as it relates to the Northern Hemisphere.

Keep in mind that seasons are relative. Summer never happens everywhere at once. Ditto fall, winter, and spring. Instead, summer occurs in one hemisphere while winter is happens in the other, and vice versa. Likewise, spring occurs in one hemisphere while autumn falls in the other, and vice versa. Thus, when the "Summer Olympics" were held in Sydney, Australia (Southern Hemisphere), it wasn't summer as far as the locals were concerned, but instead late winter.

Spring

Sometime around March 21, the vertical rays strike the equator, marking the spring equinox. This is the first day of spring in the Northern Hemisphere and the period of daylight and darkness are the same. Everyday for about the next three months, the vertical rays strike the Earth at progressively more northerly latitudes. The daylight hours get longer while night gets shorter.

Summer

On or about June 21, the vertical rays strike the Tropic of Cancer, marking the summer solstice. This is the first day of summer as well as the date that has the longest period of daylight and the shortest nighttime. From this point, the vertical rays then "turn south." Daylight hours lessen while nighttime hours increase.

Fall

On or about September 21, the vertical rays again strike the equator, marking the fall, or autumnal, equinox. This is the first day of fall and again the period of daylight and darkness are equal. The vertical rays then move into the Southern Hemisphere, striking ever more southerly latitudes each day for about the next three months. In the Northern Hemisphere, nighttime now exceeds daytime by a margin that increases each day.

Winter

December 21 is the approximate date of the winter solstice. The vertical rays then strike the Tropic of Capricorn, marking the first day of winter. Also on that date, the Northern Hemisphere experiences its longest period of darkness and shortest period of night. The vertical rays then "turn northward." For about the next three months, nighttime periods continue to exceed daytime periods in the Northern Hemisphere, but by a difference that diminishes daily. Finally, on or about March 21, the vertical rays are back at the equator, marking the spring equinox. Day and night are again equal and the seasonal cycle is complete.

Special lines of latitude revisited

Before leaving this section, discussion is in order of those other two "special lines of latitude" — the Arctic and Antarctic Circles. Each demarcates parts of the world where something peculiar happens. Specifically, every location north of the Arctic Circle and south of the Antarctic Circle experiences at least one continuous 24-hour period of daylight, and at least one continuous 24-hour period of darkness during each year. Moreover, the farther north and south one goes with respect to those two lines, the greater is the number of days of complete daylight and darkness. The extreme cases occur at the two poles, where the year is divided into one six-month-long period of daylight followed by one six-month-long period of darkness.

To help understand this, look back at Figure 9-3. The North Pole is 90 degrees' worth of latitude from the equator. On the first day of fall, the sun is directly overhead at the equator, but will appear to an observer at the North Pole to be on the horizon (90° from overhead). Everyday for the next six months, the vertical rays strike Earth below the equator. From the perspective of the North Pole, the sun is below the horizon all the while, so darkness ensues. Every other latitude between the North Pole and the Arctic Circle, Latitude 66½° North, also experiences at least one continuous 24-hour period of darkness. The significance of Latitude 66½° North is that it's exactly 90° from the Tropic of Capricorn, which marks the farthest southerly point that feels the sun's vertical rays.

Why is Christmas celebrated on December 25th?

Scripture is silent about the date of Christ's birth, which only began to be celebrated several centuries after the event. In pre-Christian Northern Europe, there was widespread belief in a sun god whose birth date was celebrated on the winter solstice. On that date, the sun is lowest in the daytime sky in the Northern Hemisphere, but rises somewhat everyday thereafter until the summer solstice. Thus, the date on which the sun began to rise symbolized the sun god's birth. As Christianity spread into those lands, mixing of religious beliefs occurred, one of which was to substitute Christ's birth for the sun god's on the winter solstice. Subsequent refinement of the calendar resulted in the few-day's lapse that now separates Christmas and the winter solstice.

The opposite occurs during spring and summer, when the North Pole has continuous daylight. As the summer solstice approaches, more and more latitudes south of the North Pole experience similar conditions. Finally, on the summer solstice, the experience of a continuous day of sunlight reaches its most southerly locale — the Arctic Circle.

Hot or Cold? Adjust Your Altitude

Altitude has an important impact on climate. The rule of thumb is that temperature and elevation are inversely related. Or in every day speech, "the higher you go, the cooler it gets and vice versa." Thus, highland areas have cooler climates than lowland areas.

Consider this example: Fewer than 200 miles separate Guayaquil and Quito, the two largest cities in Ecuador. But the annual average temperature is 77° F in Guayaquil and only 56° F in Quito. The explanation is that Guayaquil is virtually at sea level while Quito is up in the Andes at an elevation of about 9,250 feet. Despite being nearly on the equator, the city does not experience the warm climate one would normally expect at that latitude, due to the altitude factor. How does that work? Glad you asked.

Warming the atmosphere

Part of the answer concerns how the atmosphere obtains heat. A portion of the solar energy that reaches the Earth (18 percent) is absorbed directly by the atmosphere, while an even larger percentage (32 percent) reflects back into space. The largest portion by far (50 percent), however, is absorbed by the Earth, which then re-radiates that heat into the atmosphere. Thus, solar energy turns Earth's surface into something like a giant frying pan that heats the atmosphere above it. Generally, therefore, air that is at or near the Earth is relatively warm, while increasing elevations above "the frying pan" brings progressively cooler temperatures.

Weighty matter

Another reason why "the higher you go, the cooler it gets" is the fact that the atmosphere has weight. Gravity is constantly "pulling down" on it. Thus, as altitude increases, the amount of air decreases. And because the atmosphere holds heat, less air means colder temperatures. Therefore — and to return to our opening examples — Quito, which is way up in the Andes, has a colder climate than Quayquil, which is down by sea level.

Seeing (and feeling) is believing

Because relatively cool temperatures characterize high elevations, precipitation in highlands and mountainous regions is apt to be snow rather than rain for a good portion of the year — if not for all of it in the case of really high mountains. That results in what many regard as one of nature's most aesthetically pleasing sights — a snow capped mountain (as you can see in Figure 9-4). But the effects of elevation may be felt as well as seen. Go up a high mountain, and not only does it get colder as you go higher, but also breathing becomes increasingly difficult. It makes sense. Air has weight. It wants to sink towards sea level. So the higher you go, the less air is available to help warm things up and to help you breathe easier. The condition is called *thin air*.

Figure 9-4:
This snow-capped mountain demonstrates the relationship between altitude and temperature.

The lapse rate

The numerical relationship between temperature change and elevation change is called the *lapse rate*. It works out to about 3.5° F per 1,000 feet, or 6.4° C per 1,000 meters. That is, if you have two hikers on a mountain separated by 1,000 feet of vertical distance, then the person higher up is experiencing a temperature that is about 3.5° F colder than the person down slope. Parenthetically, I say "about" because *humidity* — the amount of vapor in the air — can and does tamper with these formulae. Therefore, think of the above numbers as average figures.

Windward slope, leeward slope

In addition to temperature, altitude can have a profound effect on patterns of precipitation. An air current is forced to rise when it meets a mountain (see Figure 9-5). As it does so, its vapor cools and condenses, forming raindrops (or snowflakes). Precipitation that is produced in this manner is called *orographic* (from the Greek *oros,* mountain), meaning that it is mountain-related. The heaviest rains and snows tend to occur on the *windward* slope, which is the side of the mountain from which the wind is blowing.

Figure 9-5:
Given a
dominant
wind
direction,
mountains
and
mountain
ranges may
exhibit
a wet
windward
slope and
dry leeward
slope.

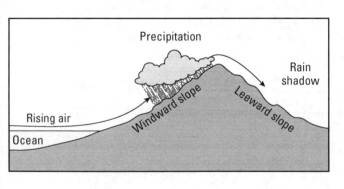

After cresting the summit, the air descends. As it does so, it warms, which is the opposite of what air needs to do in order to condense and form rain. But the air also has little vapor remaining, so the prospects for precipitation are just about zilch. The result is a dry *leeward* slope, whose droughty environs are said to be in a *rainshadow.*

Applied Geography: Locating an island resort

Many islands have distinctive windward and leeward sides. On those that do, predominant wind direction is often an important consideration in choice of location for a resort. On the windward coast the wind comes off the ocean, often resulting in rough surf conditions and blowing sand that is picked up from the beach. And if highlands are inland, rain will also result as air is forced to rise over the higher elevations. The leeward side, in contrast, is apt to experience much less rain, calmer waters, and absence of wind-blown sand. So if you're in the market for island real estate on which to build a resort, chances are good that you will opt for land on the leeward coast.

Instead of a lone mountain, as in Figure 9-5, imagine a lengthy mountain range. The windward side has a rather wet climate. The leeward side, being in the moisture-deprived rainshadow, is a desert or semi-desert. For example, the Himalayan Mountains lie perpendicular to seasonal rain-bearing winds that come from the Indian Ocean. The result is a virtual tropical rainforest climate on the southern windward side, and a vast expanse of arid and semi-arid conditions (including the Gobi and Takla Makan Deserts) on the northern leeward side.

Similarly, the Coast and Cascade Ranges in Northern California, Oregon, and Washington State, intercept moisture-bearing winds that enter the mainland from the Pacific Ocean. The result is a very moist coastal fringe (noted for its tall trees — the redwoods — and lush forests) on the windward side. But the lands to leeward are arid and semi-arid.

Gaining Heat, Losing Heat

Locations in the middle of continents tend to have hotter summers and colder winters than do locales at similar latitudes by the sea. This condition is called *continentality*. It occurs because land and water have very different characteristics when it comes to absorbing and retaining solar energy. To illustrate this point, take a look at Figure 9-6 and assume it shows a sandy beach and an adjacent lake receiving equal amounts of solar energy.

Earth and sand are not transparent, so most of the solar energy that strikes them is absorbed by and concentrated in the top-most inch or half-inch of surface material. As a result, the beach becomes super-hot. If you have ever experienced scorched feet while walking barefoot on dry sand on a sunny summer day, then you know exactly what I am talking about.

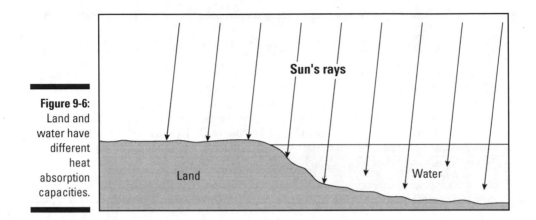

Figure 9-6:
Land and water have different heat absorption capacities.

In contrast, because the lake water has a certain transparency, solar energy penetrates the surface and, depending on the clearness of the water, spreads itself out over the depths. Also, wave action and other flow mix the upper layers of water, and thereby transport the absorbed heat away from the surface. As a result, a much greater volume of water, compared to the beach, absorbs heat. Therefore, the temperature of the lake on a sunny summer day tends to be somewhat cooler than the temperature of the beach.

This difference has significant implications for an outing at the beach, as explained below. The same is true regarding the tendency for mid-continent locales to have warmer summers and cooler winters than locations by the coast.

Afternoon versus evening

Assume it's a boiling hot summer afternoon. You and some friends are on a beach blanket at Point A on Figure 9-6. The shoreline is at Point B. You decide to take a swim. The sand is terribly hot against your feet as you run for the water and take the plunge. That water sure feels cool!

Now assume it's 9 p.m. and you are still hanging out by the lake. The sun has gone down. The temperature has cooled considerably, and the sand is now somewhat cool against your feet. Someone suggests you all go for a swim and you decide to join in, however reluctantly. So you run for the water and take the plunge. That water sure feels . . . warm!?

How can the same body of water that cools you off at midday warm you up at 9 p.m.? As we saw in the discussion of Figure 9-6, the beach gets super-hot on a sunny summer because solar is absorbed and concentrated in the top veneer of surface material. In contrast, the lake heats up more slowly because the solar energy it absorbs is spread out over a much larger volume of matter. Thus, you experience a cooling sensation when you jump into the lake in the middle of the afternoon.

But it's a different matter at 9 p.m. Because nearly all of the land's heat is contained within a thin veneer of surface sand, it tends to radiate back into the atmosphere rather fast after the sun goes down. But the lake is different. Only a small portion of its solar inventory is at the surface — exposed to the atmosphere. After the sun goes down, the lake's heat is radiated back into the atmosphere at a slower rate. Thus, it retains its absorbed energy longer than land, resulting in a warming evening swim.

Summer versus winter

The fact that land and water heat up and cool down at different rates has significant implications for climate. Look again at Figure 9-6, but instead of a beach by a lake, assume it portrays a continent next to an ocean. Also assume that Point A is a city in the middle of that continent and that Point B is a coastal city a thousand or so miles away. Time-wise, consider summer versus winter as opposed to mid-afternoon versus 9 p.m.

During the summer, the city in the middle of the continent is likely to be warmer. That is because, just like in the beach example, solar energy will concentrate at Earth's surface and heat the atmosphere overhead. In contrast, the coastal city is likely to be less warm during summer. That is because its atmosphere will be warmed by heat that radiates off both the land around it and the water offshore. But because the surface of the water contains so much less heat than the land, the total amount of heat that is radiated into the atmosphere is far less than occurs in the middle of the continent. In the parlance of climatology, the water body has a *modifying* or *mitigating effect*. That is, it results in the atmosphere being less warm than would be the case if there were land all around. The net result, whatever the vocabulary, is that the city in the middle of the continent will experience a warmer summer than the one by the sea.

Winter is another matter, however. Regarding the mid-continent city, the heat that was absorbed by the surrounding land during the summer — being concentrated at the surface and now exposed to long winter nights — radiates into the atmosphere rather rapidly, contributing to cold temperatures. These land-related conditions also apply to the coastal city. But something else of significance also affects the temperature of the atmosphere in the latter locale. Because the heat absorbed by the water body is distributed over a certain depth — and is therefore not concentrated at the surface and exposed to long winter nights — it is radiated back into the atmosphere at a much slower rate. In effect, it may serve as a source of atmospheric warmth for a significant portion of the winter, and result in a warmer winter for the coastal city.

Consider a comparison of Pierre, South Dakota and Portland, Maine. The former is in the middle of a continent, the latter is on the coast, and their latitudes are nearly the same. In Pierre, the coldest month of the year averages 17° F and the warmest month of the year averages 75° F. That makes a 58-degree *annual temperature range* (the difference between the coldest and warmest month). In Portland, the coldest month averages 23° F and the warmest month averages 68° F, making its annual average temperature range 45° F. So winters are harsher and summers are hotter in the mid-continent city.

Going with the Flow: Ocean Currents

The oceans have warm and cold surface currents that act like a global heating and air-conditioning system. They bring significant warmth to high latitude areas that would otherwise be much cooler, and significant coolness to low latitude areas that would otherwise be much warmer.

The currents also play a major role in determining the global geography of precipitation. The sun can more easily evaporate warm water than cold water, and thereby produce the atmospheric vapor that results in rain. Therefore, lands that get *sideswiped* or impacted by warm currents tend to have abundant precipitation in addition to a comparatively warm climate. Conversely, lands impacted by cold currents tend to receive very little precipitation in addition to a comparatively cool climate.

Generally, surface currents exhibit circular movements (see Figure 9-7). North of the equator, the flow is usually clockwise. South of the equator, the flow tends to be counter-clockwise. These movements are principally products of prevailing winds that "push" the ocean's surface. On the map you can see occasional exceptions to the general rules of circulation. They are the results of deflections caused by the angle at which a current strikes a land mass or the continental shelf, or by the direction of prevailing sea level winds at particular latitudes.

Warm currents, cold currents

The warm and cold portions of these circulatory systems have rather predictable geographies. As ocean currents move westward along the equator, they absorb lots of solar energy, heat up, and become warm currents. As they turn away from the equator, they generally continue to absorb about as much heat as they dissipate, at least while they remain in the Tropics — that is, the region between the Tropic of Cancer and the Tropic of Capricorn.

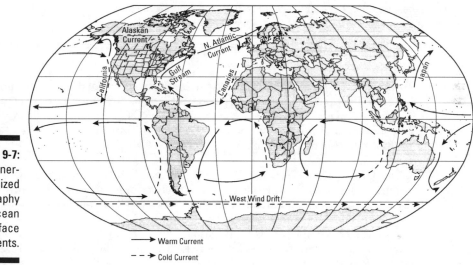

Figure 9-7:
A gener-
alized
geography
of ocean
surface
currents.

After leaving the Tropics, the reverse starts to happen: the currents radiate more heat than they gain — but slowly for the reasons you read about in the previous section concerning the ability of water to store and retain heat. Thus, the currents remain comparatively warm longer after they have left the tropics. The *Gulf Stream,* for example, is a warm-water current that moves up the Eastern coast of the United States and then becomes the North Atlantic Current (see Figure 9-7). Although it loses a fair amount of heat as it moves eastward across the mid-Atlantic, the North Atlantic Current reaches Europe with a considerable amount of stored heat remaining. As it continues to radiate that heat, it contributes to the climate of Northwestern Europe a degree of warmth that is unusual for those latitudes, and also abundant rainfall.

Take a look at an example of an area that is affected by the North Atlantic Current. Bergen, Norway (Lat. 61°N) has an annual temperature of 45° F and receives 77 inches of precipitation per year. Compare that to Churchill, Manitoba (Lat. 58° N), which, as mentioned earlier in the chapter, has an annual average temperature of 19° F and only 15 inches annual precipitation. Bergen is significantly warmer — despite its high latitude — and much wetter. The difference is partly a matter of Churchill's continentality, and partly a matter of the relatively warm current that sideswipes Bergen.

Coastal deserts

Casablanca, Morocco, highlights one of the world's most provocative geographic juxtapositions: places where oceans border deserts. Indeed, a couple of coastal deserts exist. What most have in common is a neighboring cold water current that makes evaporation difficult and rainfall unlikely.

But the Gulf Stream-North Atlantic Current is not yet finished. After impacting Western Europe, the current turns south towards the equator, now as the *Canaries Current,* to complete its circulatory cycle. By that time, however, it has lost most of the heat it once had. As a result, the Canaries Current that sideswipes Northwest Africa is quite cool.

Casablanca, Morocco (Lat. 33° N), for example, has an annual average temperature of 63° F, which is comparatively cool for a country on the fringe of the Saharan realm. It also receives only 17 inches of precipitation per year. That is a paltry sum compared to the 77 inches that the same circulatory systems dumps on Bergen, and the 46 inches of precipitation that Charleston, SC, receives by being located near the Gulf Stream directly across the Atlantic.

Going against the norm: El Niño and La Niña

You should remember that climate is an average of yearly conditions, but that in any given year very "un-average-like" events can occur. El Niño and La Niña, which happen every so many years, provide good examples. (Niño and niña mean boy and girl in Spanish.) As you can see in Figure 9-8, during an El Niño, the surface waters become unusually warm in the tropical portion of the Pacific. The reasons for this are not fully understood; but because the conditions occur around Christmas in the waters off western South America, the local populace call it El Niño, referring to the Christ child. During La Niña, the opposite happens ("girl" being the opposite of "boy") — the water is unusually cold.

Because the affected ocean water circulates, and also influences the behavior of atmospheric pressure belts (which you can read about in the next section), the impact can be substantial and widespread. Just what that means varies from place to place and year to year. Sometimes, for example, rainy seasons become extremely stormy and dry seasons become prolonged droughts. On the other hand, the effects are not always bad, as may be evidenced perhaps by a normally harsh winter that turns up mild. Generally, the media have cast "the boy" and "the girl" as climatological brats. In some times and places, however, they are the most pleasant kids you'd ever want to have around.

Figure 9-8: Conditions associated with El Niño (top) and La Niña (bottom).

Living Under Pressure

You're under pressure all the time — atmospheric pressure, that is. Just about everybody has seen a weather map with big "H's" and "L's" here and there. And just about everybody knows that they respectively stand for: high pressure and low pressure. But just about nobody understands what exactly they mean, except maybe that lows are associated with cloudy, rainy (or snowy) days, and highs usually are associated with pleasant, sunny days.

A *low-pressure system* is an area of relatively warm, moist ascending air. A *high-pressure system* is an area of relatively cool, dry descending air (see Figure 9-9). In general, therefore, you can think of low pressure as being a rainmaker, and high pressure as a drought-maker.

High pressure is so-named because the atmosphere is pressing down on the Earth. In contrast, low pressure is so-named because, due to its upward-moving air, the pressure (or weight) of the atmosphere against the Earth is comparatively low. Both are linked in a three-dimensional pattern of atmospheric circulation as shown in Figure 9-9.

Figure 9-9:
A cross-section of high- and low-atmospheric pressure-systems.

Land　　　High pressure　　　Surface wind　　　Low pressure　　　Land
　　　　　system　　　　　　　　　　　　　　system

Solar energy sets this circulatory system in motion. Some parts of Earth heat up more rapidly than others. Over those areas that do, the air tends to warm, expand, and rise. The vapor in the air cools as it rises in a *convection current,* causing condensation and (in all probability) precipitation to occur. Thus, low pressure is associated with cloudy, rainy (or snowy) conditions.

After precipitating, air at the top of a low-pressure system is cool, dry (having "lost" its moisture) and heavy. It wants to sink back down to Earth, but can't because of other air coming up from underneath. Air in the upper atmosphere therefore moves laterally until it finds a place where it can descend as a high-pressure system composed of comparatively cool, clear, and dry (low humidity) air.

Pressure belts

Because the equatorial latitudes receive a greater degree of solar energy than anyplace else on Earth, a global "belt" of low-atmospheric pressure characterizes them (see Figure 9-10). This phenomenon is called the *inter-tropical convergence zone* (ITCZ), because air from the tropics north and south of the equator is drawn into (converges on) this zone before it rises in a convection current.

The result is a warm, humid "rainmaker" that produces the tropical climates presented in Chapter 10. As implied by Figure 9-9, the air that rises in this low-pressure belt must fall to Earth elsewhere. Generally, this occurs in two sub-tropical high-pressure belts that roughly correspond to Latitudes 25-30 North and South. Given these belts of "drought-makers," it's not surprising to see desert and semi-desert conditions over much of these latitudes.

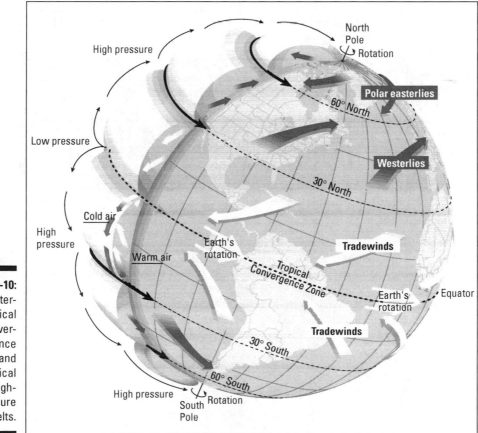

Figure 9-10: Inter-tropical convergence zone and sub-tropical high-pressure belts.

Monsoons

Now you know how and why the sun's vertical rays migrate north and south of the equator during the year. (You may wish to refer back to "Tilt-a-World: The Reasons for the Seasons," earlier in this chapter.) Because the equatorial low-pressure belt is a product of those very same rays, it migrates as well, and so do the related subtropical highs. During the year, therefore, tropical latitudes may be alternately dominated by the low-pressure belt, which brings a rainy season, and one of the high-pressure belts, which brings a dry season.

Wet season-dry season transitions happen in many parts of the tropical world, the most well-known example being the *monsoons* of South Asia. As the vertical rays move northward during summer, the low-pressure belt moves with it, drawing moisture-laden air from the Indian Ocean and producing a pronounced rainy season. During winter, the ITCZ moves south of the equator. At that time the subtropical high pressure belt moves over south Asia and surface winds blow from the interior of the continent to the Indian Ocean, resulting in a relatively rainless period (see Figure 9-11).

Figure 9-11:
Pressure belt locations during the wet and dry monsoons.

Summer Wet Monsoon   Winter Dry Monsoon

For example, Mumbai (Bombay), India (Lat. 19° N) receives 72 inches of rain during June-September. In contrast, it receives less than half an inch during December-April. For some reason, many people associate "monsoon" strictly with the rainy season. In reality, there are two monsoons, wet and dry, and each significantly impacts the region, albeit in entirely different ways.

The wettest place on Earth?

Cherrapunji, India, has the wettest recorded climate of any settlement on Earth. I include the question mark with the title because there may be wetter locales that go unrecorded. In any event, it may be near impossible to beat 425 inches of precipitation per year. That's Cherrapunji — more than an inch of rain per day on average. But even that doesn't tell the whole story. Check out the following table, and pay attention to the substantial monthly variation.

Cherrapuinji: Average Monthly Temperature and Precipitation

	J	F	M	A	M	J	J	A	S	O	N	D
Temp °F	53	55	62	65	66	68	68	69	69	66	60	55
Rain (in.)	1	2	7	26	50	106	96	70	43	20	3	1

Cherrapunji has a dry monsoon season that runs from November through February, during which it receives about 7 inches of rain. But then things change rather dramatically. In an average June and July, the town receives more than 3 inches of rain *per day,* before things taper off to a mere 2+ inches per day in August, and an inch-plus in September. Cherrapunji exemplifies the extreme conditions that can occur when two climatic determinants "pull together." In this case, a fortuitously located low-pressure system and the effects of altitude combine. The town is 4,300 feet above sea. Thus, when the wet monsoonal winds are forced to rise in and around Cherrapunji, the results are very wet indeed.

Chapter 10

From Rainforests to Ice Caps: The Geography of Climates

*T*he ancient Greeks divided the world climatically into a tropical *torrid zone,* two mid-latitude *temperate zones* (one in each hemisphere), and two high-latitude *frigid zones.* The Greeks lived in the temperate zone of the Northern Hemisphere and never sojourned to the frigid zone to their north or torrid zone to their south. That suited them just fine, for what they knew — or rather believed to be true — about those areas was fearsome.

The torrid zone in particular inspired dread. It was believed that the sun could literally burn people to death or set fire to a ship. The Greeks' direct experience with Saharan temperatures reinforced that perception, while observation of black-skinned Africans confirmed that fatal frying awaited one who ventured too close. These ideas persisted for nearly two thousand years, until 1434, when the Portuguese captain Gil Eannes rather clandestinely navigated into the area without ill effect to any of his crew.

Nowadays the study of *climate,* the average temperature and precipitation conditions that occur at a location over a long period of time, is looked upon more as a source of useful information than of fear. Knowledge of climates and their distribution help us to understand, for example, why particular patterns of agriculture are practiced in particular parts of the world. This may prove very useful in devising development scenarios aimed at increased food production. In addition, knowledge of climate helps us to understand why people live where they do (as well as where they could live), the problems and potentials of various regions, and the geographies of architecture and dress.

Obviously, therefore, climate is an immensely important geographic variable. Accordingly, *climatology* is a major sub-field of study and research within geography, and the subject of this chapter.

I'd like to throw in my own two cents here. Although each chapter is designed to stand on its own, I advise you to read Chapter 9, which recounts the reasons why particular climates occur in particular regions, prior to this one in order to grasp the full meaning of weather and climates. Global warming and climate change are important, timely, and controversial topics. While this is a logical place to discuss them, I'm going to hold off until Chapter 18, which focuses on current issues of human-environment interaction.

Giving Class to Climates

One thing that has not changed since ancient Greece is the need to classify climates. Because no two areas have exactly the same average temperature and rainfall regimes, Earth is a climatic crazy quilt. To make sense of it, various categories — climate types — have been defined on the basis of maximum and/or minimum temperature and precipitation data.

In 1898, Vladimir Köppen, a German geographer and climatologist, developed the climate classification system that is most in use today. He identified about 25 specific climates and used a rather arcane letter code (using codes such as BWh, Dfb) to identify and define them. If this book were *Climate for Dummies,* then it would be appropriate to discuss the Köppen system *ad nauseum.* But because this book's title is *Geography For Dummies,* and because I'm a nice guy and don't feel it necessary to bog you down with unnecessary information, I'm going to forego the letter code and several of Köppen's climate-types and aim for just enough descriptive treatments of just enough climates to provide a global overview consistent with the goals of this book.

Before getting down to the nitty-gritty of classifying climates, we must keep in mind a common thread among climates. Each climate type has an associated assemblage of *natural vegetation* that is likely to occur provided human beings and natural catastrophes do not interfere. Thus, if one could journey overland from the equator to the North Pole, different natural vegetation assemblages would be encountered with every change in climate. This is illustrated in Figure 10-1. In reality, of course, people have modified or eliminated natural vegetation in many areas, as by converting grasslands and rainforests to farms. When one visits a particular climatic realm, therefore, purely natural vegetation may or may not dominate the landscape, or be present at all. In any event, the climate-natural vegetation connection is so strong that some climates are named for the plant cover that is associated with them (including the tropical rainforest, steppe, and tundra.)

Figure 10-1:
The hypothetical sequence of natural vegetation in the Northern Hemisphere.

In general terms, the world's climates may be grouped into five classes. They are humid tropical climates, dry climates, humid mesothermal climates, humid microthermal climates, and polar climates. Each is discussed in the following sections.

Mixing Sun and Rain: Humid Tropical Climates

In humid tropical climates, the average temperature of each month is 64° F or higher. The warmth is a function of vertical rays (see Chapter 9) and near-vertical rays that strike the tropical latitudes pretty much throughout the year. All that sunshine, in turn, generates high evapotranspiration (for more on this, see Chapter 8), producing a moisture-laden atmosphere, and also creates convection currents (see Chapter 9 for more details) that cause the air to rise, cool, condense, and cause rain. In consequence, annual precipitation is abundant and may occur year-round or in distinctive wet seasons that vary in intensity and duration. This variation in precipitation distinguishes the three principal climates in the Humid Tropical category (see Figure 10-2): tropical rainforest, tropical monsoon, and savanna — the latter is also known as tropical wet and dry.

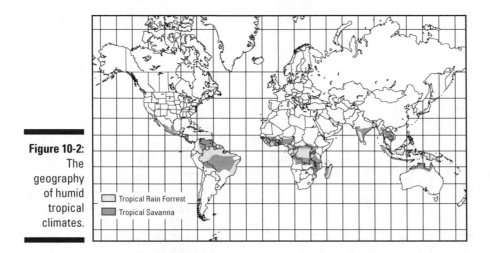

Figure 10-2:
The geography of humid tropical climates.

Tropical Rain Forrest

Tropical Savanna

As "tropical" suggests, these climates generally occur between Latitudes 231/2° North and South. Figure 10-2 shows, however, a few decidedly non-equatorial areas where a tropical humid climate prevails due to warm water currents, orographic rainfall (see Chapter 9 for more details), or some other mechanism. Conversely, "non-tropical" climates are occasionally seen between The Tropics of Cancer and Capricorn thanks to the cooling effects of elevation, cold-ocean-surface currents, or predominant wind directions.

Tropical rainforest

In areas of Tropical Rainforest climate all months average above 64° F and the driest month of the year averages above 2.4 inches (6 cm) of rain. For all intents and purposes, therefore, this climate may be described as warm and wet year-round. Although the equatorial low-pressure belt shifts north and south with the seasons, it never wanders far enough afield to result in a genuine dry period.

Incredible forests!

Tropical rainforest vegetation is the definitive natural-landscape feature. This plant assemblage is dominated by broadleaf evergreen species that grow to about 150 feet in height. Their adjoining tree tops create a "closed canopy" that in turn give rise to vertically arranged ecological zones between the ground and tree tops, each comprised of different plant species. Add to that the following:

- ✔ A year-round growing season (which accommodates a wide range of species)

- ✔ The lack of frost and drought (which also accommodates a wide range of species)

- ✔ The great age of the rainforest (which has encouraged mutation and genetic drift)

The result is the greatest concentration of living things (especially as regards to plants) to be found anywhere on Earth. How great, you ask? Well, in a square mile of forest in Vermont, you may find 12 to 15 different species of plants. In tropical rainforests, 300 to 400 different species are not unknown in comparable-size areas.

Not only is the variety of plants found here great. So, too, is their potential as sources of food and medicine. The latter is particularly important because a high percentage of medicinal drugs — including anti-carcinogens — utilize chemical compounds derived from tropical plants. Thus far, however, only a relatively small percentage of rainforest plants have been analyzed for their food and medicinal values. Also, because plants take in carbon dioxide and give off oxygen, the tropical forests are important to maintaining global atmospheric balance.

Endangered forests

Despite their real and potential benefits, tropical rainforests have been disappearing at an alarming rate. Reasons include:

- ✔ Rapidly rising populations in rainforest countries, which encourage conversion of forests to farmlands

- ✔ Global demand for timber coupled with technological developments that make rainforests more harvestable than ever before

- ✔ Government programs that encourage settlement of rainforests either to assert ownership of remote areas or to relieve population pressure in other parts of the country

Fortunately, countries that contain rainforests have created national parks and preserves that will save millions of acres for posterity. But millions of acres more stand to be lost unless preservation efforts are greatly expanded.

Poor soils

Despite the lush forest cover, the tropical rainforest realm is underlain by infertile soils called *latosols*. These are products of warm temperatures and high rainfall, which respectively encourage high microbial activity that breaks down topsoil nutrients, and wash them away (a process called *leaching*) by means of runoff or downward percolation of water through the soil. Either way, the effect on soil nutrients is much like what happens to the contents of a tea bag after it has been used a couple of times — it becomes weak.

If the soil is so bad, then why (you may ask) is the natural plant cover so lush? The answer is found in root systems that tend to fan out laterally from the bases of plants rather than dig vertically into the soil. This allows trees to effectively absorb nutrients in the topsoil before leaching does its thing.

Shifting cultivation

A large percentage of the people who live in rainforest countries farm for a living, many of who practice *shifting cultivation,* which has a particularly devastating effect on rainforests. In this form of agriculture, farmers (and their extended families) clear an area of forest, grow crops on the plot for a year or two, and then abandon it, only to move on (hence, shift) to a new area of forest and repeat the process.

Soil infertility explains this practice. When farmers remove the trees, they also remove the sources of leaf-fall that contribute to productive topsoil. With the nutrient source literally cut off and the soil exposed to direct sunlight and rainfall, leaching is swift and sure.

When a plot of land has been abandoned, the forest reclaims it and the fertility of the soil gradually improves. After lying *fallow* (plowed land that's not being farmed) for a number of years, it may be used again. But population growth in most rainforest countries is so high that "recycled" land alone is insufficient to meet local food needs. As a result, new areas of virgin rainforest must be annually cut down and the acreage added to the inventory of land that is used for occasional shifting cultivation.

Tropical monsoon

Unlike the rainforest realm, tropical monsoon regions experience a distinct dry season. That is because tropical monsoon regions generally lie in the area located between Latitudes 5° and 10° North and South, where the drying effect of the sub-tropical high-pressure belts are felt for part of the year. Despite high annual rainfall, the periodic dryness is sufficient to prevent the presence of many plant species that grow under tropical rainforest climatic conditions. Other than that, the characteristics and problems facing the world's monsoon lands are rather similar to the rainforest realm. In fact, expect for the monsoon's wet and dry seasons, the characteristics are so similar that many climate maps include tropical monsoon regions within the tropical rainforest category — as is done in Figure 10-2.

Malaria: A case study in climate and disease

Each year about 300 to 500 million cases of malaria occur globally and some 1.1 million people — mainly residents of sub-Saharan Africa — die from the disease or related complications. The disease is caused by a parasite that consumes red blood cells, causing high fever and other side effects. Species of mosquitoes that belong to the genus *Anopheles* are responsible for malaria in humans.

When a mosquito "bites," it actually sticks a syringe-like appendage into its victim and sucks up some blood. (Mosquitoes have no teeth, so they literally couldn't bite if their lives depended on it.) This rather antisocial behavior is necessary for mosquito reproduction, blood being required to produce eggs. Accordingly, and with no disrespect intended, it is only the females that bite. When a mosquito sucks up blood from a person or animal that has malaria, it may also suck up the malarial parasite. Hundreds of species of mosquitoes exist; and fortunately for humans, in just about every case, sucked-up malarial parasites die soon after entering the insect's body. For whatever reason, however, in the body of an *Anopheles* mosquito, the parasite remains viable, and thus is capable of being spread to the next human or animal that the insect "bites." The geography of malaria, therefore, is largely determined by the geography of the *Anopheles* mosquito, which is in turn is determined by the geography of the environmental conditions that the mosquito requires in order to live and reproduce. The principal criteria are temperatures that stay above about 75° F, and standing, shaded fresh water in which the insect can lay its eggs. In other words, it requires conditions that exist in abundance in areas that experience humid tropical climates.

People used to think that tropical air was unhealthy (hence, *mal aria* — "bad air"), and in many circles, belief persists that tropical climates are dangerous to humans. In truth, nothing is innately harmful about a warm and humid atmosphere. But what does occasionally happen is that climate gives rise to environmental conditions that are ideal for the proliferation of an insect or critter that is instrumental to the spread of a particular disease. The connection between humid tropical climates and malaria is a case in point.

Savanna (tropical wet and dry)

Savanna climate is distinguished from its tropical climatic kin by pronounced wet and dry seasons. The rather even duration and importance of these seasons have given rise to a self-explanatory climatic alias, tropical wet and dry. Most of the savanna realm is located between latitudes 5° and 20° North and South. These regions are alternately affected by passing low- and high-atmospheric pressure belts, which bring with them the wet and dry seasons respectively.

"Savanna" refers to the natural vegetation that occurs under these conditions: a mix of trees and grasses (see Figure 10-1). The relative abundance of these elements generally varies, however, with annual precipitation. Accordingly, trees dominate where rainy seasons are relatively long. Grasses dominate under opposing conditions. In some parts of the world — Africa in particular — the grasses attract large herds of grazing animals (herbivores) and the meat-eating animals (carnivores) that prey upon them. But the grasses and relatively fertile soil that underlie them also attract herdsmen and farmers. The result, as described in Chapter 2, has been a steady decline in wild-animal habitat.

Going to Extremes: Dry Climates

"Dry climate" would seem to be a pretty straightforward concept. Wrong. Technically, it occurs where warm temperatures cause potential evaporation to exceed rainfall. Don't worry if that leaves you scratching your head. It's kind of complex, and gets worse. Herr Koeppen stipulated that the boundary between "dry" and "humid" climatic zones occurs where R < 2T + 28 when 70 percent of the rainfall occurs during . . . Like I said, it's kind of complex.

So for the sake of convenience, I choose to lose the formula and settle for the notion that a dry climate is characterized by no more than 20 inches of precipitation during the course of a year. Climatologists, true and exacting scientists that they are, may scream and rip their clothing upon reading this, but I think they will agree that the 20-inch threshold is pretty close to accurate, and entirely appropriate for the purposes of this book. The geography of dry climates (see Figure 10-3) is made up of two areas: desert and semi-dessert (or steppe).

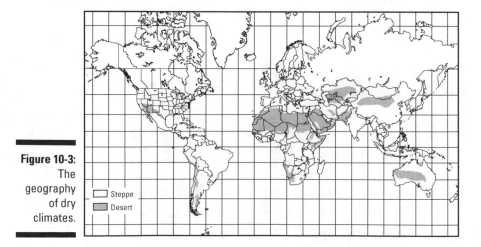

Figure 10-3:
The geography of dry climates.

Steppe
Desert

Hot times in Al-Azizia

Earth's all-time recorded high temperature — 136° F in the shade — occurred at Al-Azizia, Libya on September 13, 1922. That eclipsed the previous record, 134° F, recorded in Death Valley, California, on July 10, 1913. In all likelihood, higher temperatures have occurred on Earth but have gone unrecorded. In any event, these two numbers are testimony to the fact that by far the world's highest temperatures occur in sub-tropical deserts.

Given the discussion of sun angles in Chapter 9, this may surprise you. The equator receives higher concentrations of solar energy than Al-Azizia, so you'd figure the equator would be warmer. However, in the equatorial realm the atmosphere tends to be somewhat cloudy and contain lots of water vapor. These respectively reflect a good portion of incoming solar energy back into space and directly absorb solar energy, both before the sunshine touches Earth. Moreover, much of this area is covered by vegetation instead of bare ground, so the "Earth as frying pan" analogy simply does not work to anything near maximum efficiency at the equator.

However, the likes of Al-Azizia are a different matter. The lack of cloud cover and scarcity of vapor in the clear desert air means that a very high percentage of the solar energy that strikes this area reaches the surface, much of which is bare ground. Thus the frying pan analogy works to near perfection. The bottom line is that even though Al-Azizia receives less intense "dosages" of solar energy than does a point on the equator, it heats up to a much greater extent.

Desert

Desert climate pertains to areas that average less than 10 inches of precipitation per year. As noted in Chapter 9, cold ocean currents, persistent high-atmospheric pressure, and mountain ranges that produce *rain shadows* (leeward slopes that lack rain) create dry, desert conditions. These causative factors occur over a wide latitudinal range, which explains why deserts are rather widely distributed.

Hollywood movies have a penchant for depicting deserts as seemingly never-ending seas of sand dunes. People who live in desert areas or who have traveled through them know different, however. Most deserts are covered mainly by gravel, with enough sand and soil mixed in to support plant life. The descriptive term for this is *reg*, a word that English-speaking students of desert environments have borrowed from Arabic, a language which was born in desert surroundings and therefore has a much richer desert-related vocabulary than does English. Contrary to what Hollywood might have you believe, about 65 percent of the Sahara is *reg*. Another 30 percent is *erg*, the classic sand dune landscape. The remaining 5 percent is *hammada*, or rock-covered.

The natural vegetation of deserts consists of *xerophytes* ("dry-loving" species), which are plants that have adapted to dry conditions. To help conserve internal moisture, and thus live in lands where they would otherwise transpire to death, most xerophytes have defense mechanisms. These may include tough (even waxy) exteriors often complemented by thorns that ward off birds and other animals that might peck away at their exteriors and expose fleshy innards to the hot dry air. Because it takes a rather specialized plant to thrive in desert conditions and usually a very long time to grow to maturity in these regions of low-moisture availability, some xerophytes of the American Southwest (such as the giant saguaro cactus) are now protected by law.

Semi-desert (steppe)

Areas with semi-desert climate receive between 10 and 20 inches of precipitation per year. They normally are located between deserts and humid climate-types of either the tropical or middle latitudes. Semi-desert owns the record for the greatest latitudinal range. Instances of it are found on the equator in East Africa and in Western Canada at about Latitude 52° North. The same climatic determinants that explain the rather broad distribution of deserts also generally explain the geography of semi-desert. The natural vegetation of this climate is *steppe* — short grasses that grow in clumps with bare earth in between. Steppe is Russian in origin and describes what one sees in the vast, treeless, semi-arid plains of south-central Eurasia.

Crop-growing without irrigation in semi-desert areas is an iffy proposition. During those years when precipitation is average to above average, some production is possible. Below-average years, however, bring with them the high likelihood of crop failure and famine. Agriculturally, the steppe realm is *marginal land,* meaning that it's on the fringe (or margin) of that portion of Earth that is suitable for crops.

In contrast, the raising of livestock on the natural grasses has long been a principal activity. Accordingly, traditional pastoral nomads are associated with steppe environments, as are cowboys and cattle drives of the United States and *gauchos* (cowboy-like herdsmen) of South America. Nowadays, each of these is much more in the realm of lore than life, thanks to economic and political forces that have turned stock-raising into a rather sedentary endeavor. But raising livestock on steppe is risky. Wise resource management is essential, lest too many animals feed on the grasses, resulting in overgrazing and potential *desertification* — the conversion of non-desert lands to desert.

Applied Geography: Drought mitigation

Droughts have long been a source of human misery and death. The most horrendous ones typically occur in regions of steppe climate when the characteristic dry season is drier than usual and just won't quit. Areas to the south of the Sahara Desert have been particularly prone to these occurrences in recent decades. Adding to their devastation is the relative remoteness of this region, which hampers awareness of the drought in the outside world, thus inhibiting the ability of relief agencies to mount an effective response.

Thanks to *remote sensing,* the use of satellite imagery to monitor Earth's surface, it's now possible to monitor the onset of drought as it happens. This is made possible by satellite-based infrared imaging, which is somewhat like picture taking. Lush, healthy vegetation has lots of chlorophyll, which is an excellent reflector of the infrared energy that is a part of sunshine. In color infrared imaging, such vegetation registers as bright red. In contrast, dry vegetation (which is low on chlorophyll) appears as brown. Thus, when the dry season ends and the rainy season begins, the landscape rather immediately changes from brown to red, at least as far as an infrared sensor is concerned. If however, brown persists, then that means that the rains have not arrived, possibly indicating the onset of drought.

Enjoying the In-between: Humid Mesothermal Climates

Humid *mesothermal* (moderate temperature) climates are located in the low-middle latitudes. They typically receive more than 20 inches of precipitation per year and therefore are not "dry." Also, the coldest month is less than 64° F but above 27° F, which places these climates between the tropical and polar temperature thresholds (see Figure 10-4). Mild winters and natural vegetation that is dominated by deciduous trees (which shed their leaves annually, as opposed to evergreens) are definitive characteristics of these areas, which include the Humid Subtropical, Mediterranean, and Marine West Coast climates.

Humid subtropical

This climate type is characterized by year-round precipitation, warm summers, and cool winters. Much of the Southeastern United States is in this category, which helps explain why balmier parts of that region are favored nowadays by so many people as a desirable place to live and retire. Other parts of the world that experience this climate include Southern China and the sub-Himalayan lands, Southeastern South America, and parts of eastern Australia.

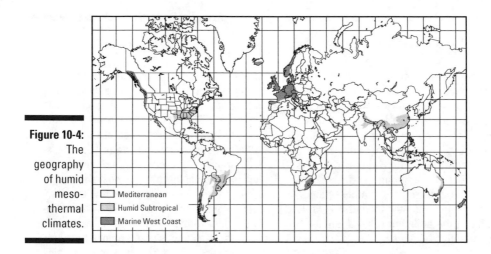

Figure 10-4:
The geography of humid meso-thermal climates.

The short winter of this climate-type makes for a long *growing season,* the average number of days between the last frost of spring and the first frost of fall. As a result, agriculture tends to be an important activity. In the lower latitudes, where freezing temperatures are rare, citrus and other frost-sensitive tree crops may be plentiful. But most anything can grow here to good effect. That includes a majority of the Earth's rice crop (mainly in Asia), arguably the world's most important staple food.

Mediterranean

The principal characteristic of this climate is its dry summer. Though found on all continents, Mediterranean climate is most associated with — you guessed it — the land around the Mediterranean Sea. The natural vegetation consists of grasses and scrubs of various sorts, which may become a tinder-box during the dry season. Much of California (save the mountains, deserts, and northern coast) experiences this climate, which explains why wildfires are a common dry-season hazard in that state.

Various fruits prized for their sweetness and/or juices (grapes in particular) dominate agricultural land use. Lack of rain during summer, when the fruits ripen, deprives plants of moisture that would be taken up by root systems and dilute the natural juices. This is a very big deal for the wine industry, for which valuable vintages have everything to do with lack of rain while grapes mature. As to why this kind of climate happens in the first place, predominant wind systems (associated with shifting atmospheric pressure belts) blowing from the land to the sea during summer is the culprit.

Marine west coast

This climate is characterized by mild to cool summers and cool winters. Its name pretty much tells you where you'll find it — on the west coast of continents in the middle to high-middle latitudes. For virtually every occurrence, the principal determining factor is a warm ocean surface current immediately offshore. Coastal mountain ranges prevent this climate from spreading inland and occupying large areas in North America, South America, and Australia. Conversely, lack of Atlantic coastal mountains in much of Europe allows the marine atmosphere to uninterruptedly waft eastward and characterize most of that region.

The natural vegetation of this climate is mixed coniferous (needle-leaf) and deciduous forests, the two types respectively dominating in the higher and lower latitudes. Over the centuries humans have deforested much of that portion of Europe where this climate is found and converted the land to agricultural use — the warm temperatures attendant to the warm currents serving as a boon for agriculture. The same is generally true of the marine west coast areas of Australia, New Zealand, and South Africa.

In North America, relatively little conversion of forests to farms occurred in areas of marine west coast climate because of the prevalence of mountainous terrain. As a result, the natural vegetation serves as the preferred crop — which is to say that forestry is a major endeavor. And what forests they are! The relatively high temperatures and abundant precipitation wrought by the warm-water current offshore created lush and majestic stands, including the redwoods of Northern California. As demand for timber — much of it emanating from Asian markets — has grown, so has controversy between loggers and environmentalists regarding the extent and nature of cutting. The future of the relatively few remaining "old-growth forests" is a particularly contentious flash point.

Cooling Off: Humid Microthermal Climates

Humid *microthermal* (low temperature) climates are found in the high-middle latitudes. Sun angles are rather low in these areas. As a result, the average temperature of the warmest month only surpasses 50° F while the average temperature of the coldest month is 27° F or less. The two principal climates in this group are humid continental and subarctic (see Figure 10-5). Summertime differences distinguish the two climates. In *humid continental* regions at least four months of the year average above 50° F. In *subarctic* climatic regions, in contrast, fewer than four months average above 50° F. Forest is the dominant natural vegetation in both areas. Lack of land in the high latitudes explains the absence of these climates in the Southern Hemisphere.

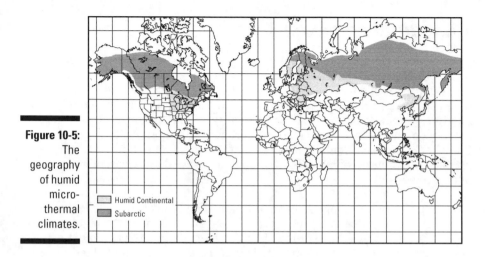

Figure 10-5:
The geography of humid micro-thermal climates.

Humid continental

This climate is found principally in Northeastern China, Eastern Europe, and, in North America, the Northeastern and Upper Midwestern parts of the United States and adjacent areas of Canada. Over much of these lands, the natural vegetation has given way to farmland. In the United States, dairy farming and the corn-soy complex (popularly called *the corn belt*) dominate the more humid east, while wheat and other hardy grains dominate the drier west. Much of the more northerly part of this realm is a bit too cool for agriculture, so forestry is intact. Coniferous softwoods, highly prized sources of pulp, dominate and support locally important logging economies.

Subarctic

This climate is generally found immediately north of the humid continental realm. Temperatures are too cold for too long for deciduous trees to thrive, and therefore coniferous forest (called *taiga,* a word of Russian origin) dominates the natural vegetation. Indeed, for the most part these forests are intact because the same chilly climes that discourage deciduous tree-growth also preclude agriculture. As a result, the broad belt of subarctic climate that extends all across the northerly portions of North America and Eurasia represents the largest expanse of forest on Earth. Generally, however, these forests are well removed from markets and mills and are therefore relatively untapped. On the whole, the subarctic realm is lightly populated. Mining is locally important and accounts for most towns' economies.

Vertical zonation and "highlands climate"

Vertical zonation refers to the changes in climatic conditions and their associated vegetation that are observed between the base of a high mountain and its summit. To take an example from East Africa, the base of Mount Kilamanjaro lies in tropical savanna climate. Its summit, however, which is 19,430 feet above sea level, is covered partly by snow and ice. For all intents and purposes, therefore, a hike from the base to the summit is tantamount to traveling from the tropics to the poles, experiencing enroute climate and vegetation change that would normally require a journey of several thousand miles. On some world climate maps, mountain ranges are shown as having *highlands climate*. This refers to the presence of the multiple climates of vertical zonation, instead of a singular climate type that is unique to mountains.

Dropping Below Freezing: Polar Climates

Cold temperatures are the dominant characteristic of polar climates. The average temperature of the warmest month is less than 50° F, and most months typically average below freezing. The very small doses of solar energy that occur at these polar latitudes, despite the long daylight hours of summer, explain the frigidity. The resulting natural vegetation (if any) consists of short grasses, mosses, lichens, and an occasional stunted tree or shrub. The two climates that make up the geography of polar climates (see Figure 10-6) are tundra and ice cap.

Figure 10-6: The geography of polar climates.

Tundra

Ice Cap

Tundra

In areas with *tundra* climate, at least one month of the year averages above freezing (32° F), but not above 50° F. Like the humid microthermal climates, and for the same reason, tundra is almost exclusively a Northern Hemisphere phenomenon (small parts of Antarctica experience it). *Tundra* is a Russian word that refers to the vast, nearly treeless landscapes that are characteristic of this climatic region.

Lack of forest is not a function of cold air temperatures *per se,* but rather the frozen soil that persists for nearly the entire year, and which prohibits tree roots from taking in sufficient nutrients. Thus, grasses that grow in abundance during the long daylight hours of the short and chilly summer dominate the natural landscape. This vegetation, in turn, attracts huge herds of caribou that annually migrate to the tundra to feast and fatten up for the long winter ahead. This relationship between plant and animal is the principal reason why, in Alaska at least, large portions of the tundra region have been designated as National Parks or National Wildlife Refuges.

Because the growing season is so short, agriculture is virtually unknown. Thus, the livelihood of the traditional societies who have long inhabited this realm — the *Inuit* (formerly known as Eskimos), and neighboring Native Americans in the Western Hemisphere, and the Lapps and neighboring peoples of the Russian Arctic, have relied on hunting or herding. In the last couple of decades, however, the economic importance of the tundra has increased dramatically as a result of the discovery of significant quantities of petroleum and natural gas. Serious and sometimes acrimonious debate has resulted, pitting proponents of resource exploitation and pristine wilderness.

Keeping permafrost frozen

A peculiar environmental phenomenon that has a powerful and direct bearing on debate between drilling for natural resources and providing for wildlife is *permafrost,* permanently frozen soil that underlies the tundra. If you drill for oil in the tundra and send it through a pipeline to wherever, then the oil needs to be warm. That is because crude oil, which is thick and viscous, flows very haltingly through cold pipe. And cold pipe is something that Alaska's climate virtually guarantees for most of the year. Fortunately, crude oil is hot as it comes out of the ground, and the warmth helps make it less viscous as it flows through the pipe. But the warm oil warms the pipe, which can melt the permafrost underneath, causing the pipe to sink into the ground and break, resulting in oil spills. For that reason, and at great expense, the Trans-Alaska Pipeline is elevated on stanchions for much of its route. This and other permafrost-related problems are at the heart of debate concerning possible future oil exploration and drilling in the tundra.

Ice caps: Hollywood-style

Virtually every movie you have ever seen about the Arctic or Antarctic contains the obligatory blizzard scene in which the attendant "white out" easily lends the impression of copious snowfall. Indeed, big snowfalls may occasionally happen, but ice cap blizzards more typically result from high winds kicking up loose snow that never melted and has yet to become consolidated with the ice cap.

Ice cap

In areas of ice cap climate, every month averages below freezing. As the name suggests, its distribution generally coincides with the ice caps that overlie Greenland and Antarctica. At low temperature air can contain very little water. Also, due to the cold temperatures, relatively little evaporation takes place in the polar realm. On both accounts, therefore, the air has a rather low supply of vapor, which in turn depresses the possibility of precipitation. Thus, ice cap climate technically qualifies as desert because it receives less than 10 inches of precipitation per year. The astonishing and uniquely low temperatures, however, result in its being granted its separate climatic status.

As noted in Chapter 8, the ice caps in Greenland and Antarctica are a couple of miles thick, and all of it is the result of precipitation, which would seem to contradict what I just wrote. But all of that ice, however, is the product of thousands of years' worth of small annual accumulations of snow, which, due to the year-round cold temperatures, tend not to melt, but instead accumulate, compact, and add to the ice cap.

These facts mean that boring down into the ice cap is rather like going back in time. The deeper you bore, the older the ice — and thus the older the precipitation (snow) of which the ice is made. Moreover, because precipitation captures minute but measurable amounts of atmospheric gases, it constitutes a record of the nature of the atmosphere at the time that it fell. Thus, the ice caps are important sources of information about the atmosphere and climates long ago, which figures prominently in contemporary inquiry concerning global warming and its causes.

Part III
Peopling the Planet

"Theory has it the island was once a part of Manhattan."

In this part . . .

Where have you gone, Adam and Eve? Life was so simple then. Two people. One culture. Nice garden. Plenty of room for growth. If somebody else had been around to write about population geography, I'm guessing three paragraphs would have sufficed. Not any more.

Today more than 6 billion people are divided into who knows how many thousands of cultures. And these folks just won't stay put. Ever since the original twosome got their eviction notice, people have been moving and migrating, rendering population geography into something akin to a restless tide. In the midst of change, however, discernible patterns (constants, if you will) emerge that concern the geographies of population, culture, migration, and control of the planet.

In this part, you will learn some of the key concepts and concerns of human geography. And yes, it takes more than three paragraphs. Indeed, it consumes four chapters that address the topics just mentioned — population, culture, migration, and control of the planet. Even that doesn't complete the story, for we still have the matter of how people use and misuse the planet. Stay tuned.

Chapter 11

Nobody Here but Us Six Billion

*N*ot long ago, the global population — the number of people worldwide — passed the 6 billion mark. That number has little meaning by itself. But if you consider that 200 years ago the global population was "only" 1 billion, then today's total gets your attention pretty quickly. A logical first reaction is that the birds and the bees have been working overtime. Indeed, those little critters have a certain way about them. But global population trends involve more than what happens in the privacy of a nest or hive.

The pages ahead focus on *population geography*, which analyzes the distribution of people and their characteristics over the face of the Earth. Of necessity, this involves a smattering of *demography*, the science of vital statistics. "Vital" refers here to life, as when medical equipment is used to monitor a patient's "vital signs." Thus, demography involves birth rates, death rates, life expectancy, and other numerical indicators of the human condition.

For people who love to calculate statistics, demography is a dream come true. Chances are good, however, that you are not one of those people. So I forego the arithmetic and focus on generalizations and implications that result from it. Most of all, I focus on how humans and some of their vital attributes vary geographically.

Migration is an important factor in population change both internationally and within individual countries. Indeed, I am going to hold off on that subject for now and instead devote the entire next chapter to it because of its importance.

Going by the Numbers

The world's 6.1 billion people are spread very unevenly across the planet's surface, as you can see from Figure 11-1. Virtual empty quarters — large and totally uninhabited realms — correspond with the ice caps and tundra of Antarctica, Greenland, and the very high latitudes in general. Similarly, large desert areas often are low on people. Indeed, if you read the chapter on climate (Chapter 10), then it should come as no surprise that the Sahara, Gobi, Arabian, and other desert realms are fairly devoid of people. Also, most of the world's rainforest realms have low population densities, as the Brazilian interior and central Congo indicate.

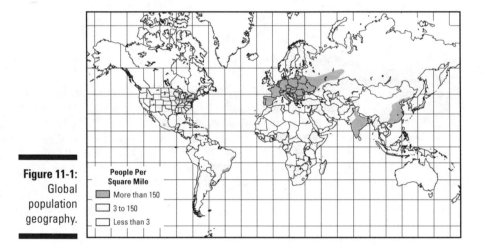

Figure 11-1:
Global
population
geography.

People Per
Square Mile

More than 150

3 to 150

Less than 3

But for every desolate area, you must consider the likes of Hong Kong, with some 16,000 people per square mile, or Singapore with its 17,000 people per square mile. Those are small dots on the world map that complement large areas of comparatively high density: the northeastern U.S. and adjoining areas of Canada; much of Western and Central Europe; the Nile valley; north central India; eastern China, and Japan and Java.

Table 11-1 on the world's most populous countries highlights the dominance of China and India, which respectively are home to 21 percent and 17 percent of all the people on this planet. Given those two population powerhouses, Asia contains some 60 percent of the world's population — the largest continental percentage by far (as shown in Figure 11-2). The United States is now the third most populous country on Earth, but North America as a whole contains only 8 percent of the human population. All told, the 15 most populous countries account for fully two-thirds of humanity.

Table 11-1	The 15 Most Populous Countries in 2001	
Rank	*Country*	*Population (in millions)*
1	China	1,273
2	India	1,033
3	United States	285
4	Indonesia	206
5	Brazil	172
6	Pakistan	145
7	Russia	144
8	Bangladesh	134
9	Japan	127
10	Nigeria	127
11	Mexico	100
12	Germany	82
13	Vietnam	79
14	Philippines	77
15	Egypt	70

Source: Population References Bureau.

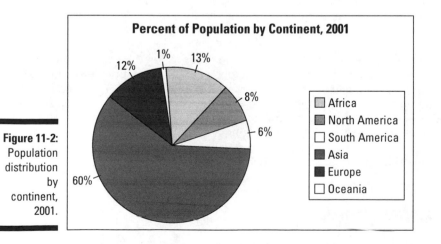

Figure 11-2:
Population
distribution
by
continent,
2001.

Dispersion versus clustering

Two areas can contain the same number of people, yet have a totally different look and feel because of the ways their populations are distributed. For example, the United States is among the minority of countries in which farmers typically live on their farms. That statement may cause you to ask, "Where else would a farmer live?" The answer is, in a village or town, and therein lies a significant difference in the way people are distributed. Rural population geography in the United States generally exhibits *dispersion,* which entails considerable open space between individual farmsteads. In contrast, the pattern in much of the rest of the world exhibits clustering. That is, farming families tend to live in a compact village, from which they walk or otherwise "commute" to the land that they tend.

The two patterns are depicted graphically in the following figures, both of which contain 21 dots that represent homesteads.

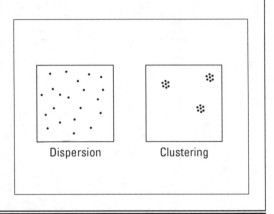

Dispersion   Clustering

Opportunity for livelihood

Trying to fully explain global population geography in a one-liner is impossible, but perhaps "opportunity for livelihood" is a good start. Human population densities tend to be high where opportunity for livelihood is favorable and low where the opposite is true. Opportunity for livelihood takes different forms and therefore, so does the characteristics of regions that support high densities.

Agricultural land in the Nile, Ganges, and Indus River Valleys, plus the valleys and coastal plains of eastern China support large populations (see Figure 11-1) due to their rich alluvial soils. How rich? Well, rich enough that since the beginning of recorded time, people who possess even the most modest agricultural technology have been able to realize sizeable harvests on relatively small acreage. In complete contrast, high densities are also found in countries where industrial and post-industrial economies dominate. Examples include the Northeastern U.S., Western Europe, and Japan.

Urban growth

Because people round the worldview cities as centers of opportunity for livelihood, one of the most significant population trends today is urban growth. I talk about that more fully in Chapters 12 and 17. But for now, you can clearly see its effects on the world population map in the likes of Greater Mexico City, the Sao Paulo–Rio de Janeiro complex in Brazil, and the major cities on the East and West Coasts of the U.S.

Globally, about 46 percent of the human race is categorized as urban, but the figures vary sharply from one continent to the next. The populations of North America, South America, and Europe are each at least 70 percent urbanized. In contrast, the urban population percentages of Asia and Africa are 47 percent and 33 percent respectively, which largely reflect continued heavy reliance on manual labor in the agricultural sector of countries' economies, plus relative lack of employment opportunities in the service and manufacturing sectors, which globally tend to be more urban-based.

Going Ballistic: Population Growth

For the vast majority of human history, total population was much, much lower than it is today. The United Nations estimates that it was only about the year 1650 that, for the first time, as many as 500 million people were alive at any one time (as illustrated by Figure 11-3). Until then, population growth could be fairly characterized as a gently rising straight line. But around 1650 something began to happen. The line started to curve upward — gently at first, but then ballistically.

Total population passed 1 billion around 1800. Thus, while it took untold eons for human numbers to reach 500 million, a mere 150 years were required to double that number. By about 1925, the figure had doubled again to about 2 billion. In the next fifty years, it doubled again, reaching 4 billion sometime around 1975. And here we are today at slightly more than 6.1 billion.

Figure 11-3:
Global population growth.

OK, everybody into Rhode Island!

Rhode Island is the smallest state in the United States. Imagine if every person on Earth went there to participate in a meeting. Would they all fit? And what would be the population density?

Rhode Island contains 1,212 square miles, and about 6.1 billion people currently live on Earth. If all of them went to Rhode Island and stood evenly spaced apart, that would work out to 5,033,003 people per square mile. One square mile (5,280 feet x 5,280 feet) equals 27,878,400 square feet. Dividing that many square feet by 5,033,003 people gives each person 5.54 square feet in which to stand. That works out to a square that is roughly 2 feet, 4 inches on each side.

Get out a foot ruler, measure an area that size on the floor, and stand in it. Now imagine being surrounded for miles and miles by people who are allotted an equal amount of space. Would you feel cramped? Well, people may respond differently to that question. But with 5.54 square feet per person, 6.1 billion people could stand in Rhode Island with little or no physical contact between them. And all of the rest of the world would be completely empty of humans.

This mathematical exercise is just that. It's not meant to play down the global impact of 6.1 billion people who need to be housed, fed, and otherwise sustained — all of which requires considerably more space than exists in Rhode Island.

Global population growth has not stopped. Instead it will continue to rise in the present century before leveling off at about 10 billion in the year 2100. Naturally, that may cause you to ask, "How can demographers be so certain about the future course of the world's population?" In fact, the experts are not certain. Instead, the future projections are (highly) educated guesses based upon reasonable assumptions concerning the global courses of birth and death.

Checking Behind the Curve: Population Change

Population change is a matter of birth, death, and/or migration. That is, in a given year in a given country, some people are born, some people die, some people move in, and some people leave. Demographers have developed a statistically based vocabulary that addresses these issues. Three terms in particular are worth passing along to you at the present time because they appear frequently in the following pages:

✔ **Birth rate:** The annual number of births per 1,000 population.

✔ **Death rate:** The annual number of deaths per 1,000 population.

✔ **Natural increase:** The annual rate of population change as calculated by subtracting the death rate from the birth rate. (Typically, the birth rate exceeds the death rate, so population rises. But occasional short-term calamity such as a plague, war, or economic turmoil may produce the opposite effect.)

Dealing with births and deaths: Natural increase

Just as humans are unevenly spread across the surface of the earth, so, too, is population growth. Indeed, perhaps the single most important demographic reality of our times is that the rate of natural increase differs dramatically in different countries and regions of the world (see Figure 11-4). The highest rates tend to be found in Africa, South Asia, and Latin America. The lowest rates occur in North America, Europe, northern Asia, plus Japan, Australia, and New Zealand.

On the map, a "high" rate of natural increase is considered to be in excess of 2 percent. That may not seem like a lot, a 2 percent rate of natural increase will double a country's population in 35 years. Therefore, every country in the "high" category on the map doubles its citizens in some time less than that. Indeed, approximately half of the "high" countries have rates of natural increase in excess of 2.5 percent (a doubling time of 28 years) and about half of those have rates of 3.0 percent (23 years) or greater.

Natural Increase, 2001

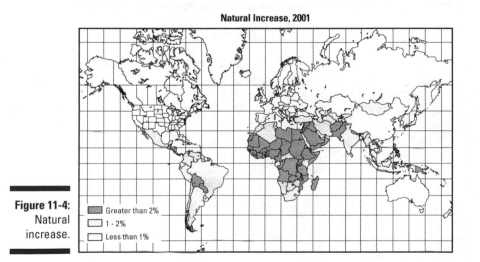

Figure 11-4: Natural increase.

- ▓ Greater than 2%
- ☐ 1 - 2%
- ☐ Less than 1%

Rapid growth, poor country

Rates of natural increase and their associated "doubling times" (see the "On the double!" sidebar) have incredible implications for economic development and social well-being. A 2.5 percent rate of natural increase means there will be about twice as many mouths to feed in 28 years; twice as many people in the schools; twice as many people seeking health care, energy, employment, and transportation. Now, it would be one thing if this pertained to affluent countries that could adequately meet the needs of their citizens. But what about a poor country?

And indeed, that's the rub. Generally, countries that exhibit the highest rates of natural increase are relatively poor. Stated differently, the highest rates of population growth are generally taking place in countries with the least amount of financial resources to address the needs of their growing population.

Slow growth, affluent country (with some notable exceptions)

Figure 11-4 includes a "low" category for which the rate of natural increase is under 1.0 percent. That corresponds to a population "doubling time" of 70 years, so every country in this category takes at least that long to double its present citizenry (migration not considered). Often, low rates of natural increase coincide with relatively affluent countries. A principal reason is that the economies and livelihoods of well-to-do countries do not require large amounts of manual labor, so families have comparatively few children, and that depresses the rate of natural increase. (Additional reasons are discussed in the next section.)

On the double!

How long does it take a country to double its population given a particular rate of natural increase? The following table provides some answers. To take two examples, if a country has a 1.5 percent rate of natural increase, then it will take 48 years to double its population, assuming the rate of natural increase does not change in the interim. In contrast, a 3.0 percent rate of natural increase is sufficient to double the population in only 23 years. Remember that natural increase equals birth rate minus death rate. It does not take into account either in-migration or out-migration, both of which could be key factors in a country's population change.

Rate of Natural Increase (percent)	Doubling Time (years)
0.5	141
1.0	70
1.5	48
2.0	35
2.5	28
3.0	23
3.5	20

However, many countries have doubling times far longer than 70 years, and some even have negative rates of natural increase. Germany, for example, has a natural increase of –0.1 percent. Russia's figure is –0.7 percent. That means the death rate is actually exceeding the birth rate in those countries, whose populations may begin to decline should these conditions hold for the foreseeable future.

Most examples of "negative growth" are former Communist countries that are experiencing economic difficulties as they struggle to make the transition to market economies. One way to economize in tough times is not to have children, which is why those countries' birth rates have dropped below their death rates — negative growth.

Increasing for a reason: The demographic transition model

The geography of natural increase doesn't "just happen." Instead, particular rates of natural increase are occurring in particular countries for particular reasons. To help explain these circumstances, demographers have developed a widely applicable set of generalizations (based on the experiences of many countries) called the *demographic transition model,* which is shown in Figure 11-5. Because the topic is people, *demographic* makes complete sense. Also, the model begins and ends with nominal population change. But in between a period of *transition,* characterized by substantial growth, occurs.

The demographic transition model considers the relationship between birth rates and death rates over time, and consists of four stages:

Stage 1: High stationary

Stage 2: Early expanding

Stage 3: Late expanding

Stage 4: Low stationary

I am going to make a big deal of them for two reasons. First, if you understand the model, then the factors that gave rise to the historical population growth curve illustrated in Figure 11-3 become crystal clear. Second, different countries of the world are in different stages of the demographic transition. Therefore, if you understand the model, then you can understand why a particular country is experiencing its particular rate of natural increase and its attendant social characteristics. Parenthetically, several developed and developing countries have already completed the transition. Demographers have used the experiences of those countries to predict the course of global population growth into the next century (see Figure 11-3).

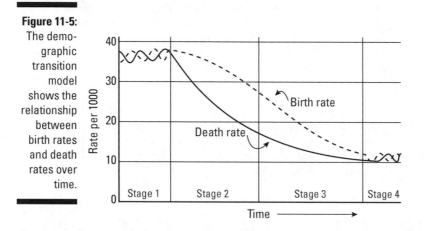

Figure 11-5:
The demographic transition model shows the relationship between birth rates and death rates over time.

Rate per 1000

40

30

20

10

0

Birth rate

Death rate

Stage 1   Stage 2   Stage 3   Stage 4

Time

Stage 1: High stationary

In this stage, birth rates and death rates are *high* and about equal (see Figure 11-5). Thus, population growth is *stationary* (exhibiting little or no change) because the numbers of births and deaths are "canceling out" each other. These conditions were characteristic of global population prior to 1650 (Figure 11-3). The high death rates of those times were products of poor sanitation, tainted water supply systems, faulty food storage, lack of education, and absence of medicines and vaccines. The results included the following:

 ✔ *Infant mortality* (the incidence of death before a child's first birthday) was astonishingly high, meaning that it claimed about 25 percent or more of newborns even in some "advanced" societies.

 ✔ *Average life expectancy* (the number of years a newborn could expect to live) was low. How low? Well, today the average citizen of France can expect to live 78 years. But church and cemetery records suggest that in the 1600s, French life expectancy was about 35 years.

   On average, therefore, people died young. Many never reached reproductive age, and those who did tended not to live that many years during their fertile time of life.

Human societies have typically responded to high death rates by having high birth rates, and the time prior to 1650 was no exception. Factors that contributed to high birth rates included the following:

 ✔ Most families farmed for a living, so more children meant more hands to do the manual labor that was necessary in those days before widespread use of machinery.

 ✔ Retirement pensions, 401(k)s, life insurance, and social security checks were unknown. Having children (and the more, the better) guaranteed there would be somebody to look after you if you were fortunate enough to reach old age.

✔ Given short life expectancy and need for children, people — especially females — married young. Most women were wed by their mid-to-late teens and, not withstanding the often-fatal rigors of childbirth, had been through a couple of pregnancies by age 20.

Virtually no country currently experiences this range of conditions in its entirety. Nevertheless, an understanding of these circumstances is very important because they set the stage for other phases that are very real in the present age.

Stage 2: Early expanding

Birth rates exceed death rates by a widening margin in this stage (see Figure 11-5). When that happens, population does more than simply grow: It increases dramatically. Look again at Figure 11-3 and note the S-shaped curve of population growth. The conditions just described correspond to the bottom — that is, *early* — half of the curve, when population was *expanding* at faster and faster rates after years of being stationary.

Hence the name of this stage is *early expanding,* which nevertheless perplexes many people because, as you can see on Figure 11-5, birth rates and death rates are declining throughout it. The key thing to focus on in that diagram is the widening gap between birth rates and death rates that is characteristic of this stage. Even though both rates are dropping, the gap between them is widening, birth rates being the higher of the two. Thus, population grows (expands).

But why are the rates dropping and the gap widening? Basically, birth rates drop because of a tempering of the last bulleted items of Stage 1, but death rates are declining much faster because of the following:

✔ Basic and widespread improvement in water supply and sanitation are having a very positive effect on public health.

✔ Medicines and vaccines are becoming accessible to more and more people.

✔ Infant mortality is dropping and life expectancy is rising. More people are reaching reproductive age and are reproducing. People are living longer in their reproductive years and are reproducing more.

✔ Death rates are dropping faster than birth rates, so population grows — slowly at first, and more dramatically more recently.

The widening gap between birth rates and death rates results in growing rates of natural increase. Thus, looking again at the global map of natural increase (see Figure 11-4), most countries in the "high" category are in the early expanding stage.

Stage 3: Late expanding

Birth rates exceed death rates by a narrowing margin in this stage (see Figure 11-5). When that happens, population grows but at rates that are progressively slowing. Look once at Figure 11-3 and its S-shaped curve of population growth. The conditions just described correspond to the top — that is, *late* — half of the curve, when population was *expanding* at slower and slower rates after years of skyrocketing. Hence the name of this stage is *late expanding*. Because birth rates exceed death rates by a *decreasing* margin, the result is a lowering of the rate of natural increase. On the map of natural increase (see Figure 11-4), most countries in the "medium" category are experiencing this stage and its attendant social conditions, which include the following:

- ✔ Improvements continue to be made in public health, resulting in lower infant mortality and longer life spans. Thus, the death rate continues to decline.

- ✔ As the economy develops, machines perform increasing amounts of work that used to be done manually. Thus, the incentive to produce children strictly for their labor potential drops.

- ✔ More people gain work in jobs that provide pensions and retirement systems. This lessens another historic incentive to produce children.

- ✔ Increasing numbers of women encounter career and educational opportunities that have the effect of delaying marriage and child-bearing.

- ✔ Increasingly, husbands and wives consider the costs of raising and educating children and opt to limit the size of their families.

As a result of the last four factors, the birth rate begins to decline — slowly at first, but then more rapidly as the modern economy encompasses more and more families. As the gap between death rate and the birth rate diminishes, the rate of population increase begins to slow, and the curve exhibits signs of leveling off.

Stage 4: Low stationary

In this final stage birth rates and death rates are *low* and about equal (see Figure 11-5). Thus, population growth is *stationary* (exhibiting little or no change) because the numbers of births and deaths are "canceling out" each other. On the map of natural increase (see Figure 11-4), most countries in the "low" category are experiencing this stage. (Global population as a whole, as per Figure 11-3, will probably not experience this stage until early in the next century.) The characteristics of the low stationary stage are as follows:

- ✔ Continued improvement and increased availability of health care results in continued lowering of the death rate.

- ✔ The economy is overwhelmingly industrial or post-industrial, resulting in diminished need for manual labor except in those remaining occupations that require a high degree of craftsmanship.

✔ Institutional retirement systems and benefits are widespread, nullifying the need to have children for the sake of social security.

✔ More women defer marriage and motherhood (or opt out entirely) as educational and career opportunities become more widely available and socially acceptable. The effects lower the birth rate.

✔ Average family size continues to decrease as more families factor in the costs of child raising and educating children.

Making connections

In each stage of the demographic transition model (see the previous section), natural increase is closely related to other demographic variables, each of which can be mapped and analyzed, and thus reveal a broader appreciation of the geography of the human condition. The following sections offer maps and brief discussions of three variables that illustrate the possibilities.

Wealth (Gross National Income [GNI] per capita)

A map of global wealth reveals that the world's most affluent countries are found in North America, Western Europe, and selected "outlying" places such as Japan, Australia, and New Zealand (illustrated in Figure 11-6). At the other extreme are numerous countries in Africa and Asia. If the overall pattern of rich countries and poor countries on this map looks vaguely familiar, it ought to. As suggested earlier and confirmed here, you can see an inverse relationship between natural increase and wealth. That is, countries that have a high rate of natural increase generally have low average wealth, and vice versa. And again, the most significant implication is this: The highest rates of natural increase are occurring in countries that have the least financial means to see to the needs of their rapidly expanding populations.

Percent of population under 15 years of age

Figure 11-7 — a world map showing percent of population under 15 years of age (by country) — reveals a familiar pattern. The highest category on the map pertains to countries in which more than 40 percent of the population is in that age category. And basically, those countries are found in Africa, plus Central America and Southern Asia. The lowest rates, in contrast, tend to be found in North America, Europe, and East Asia, plus Australia and New Zealand. Thus, the highest percentages of young people tend to be found in countries that have rapidly expanding populations and the least financial wherewithal that can be brought to bear on the education, health, and nourishment of the next generation.

Infant mortality

Infant mortality is a sensitive indicator of public health and education. Figure 11-8 shows the highest rates of infant mortality are occurring in countries that have the highest rates of natural increase, which also tend to be poor.

That lack of wealth is largely responsible for the poor states of health care and sanitation that produce high rates of infant death. Co-occurrence of high rates of infant mortality with high rates of people under 15 years of age suggests a societal preference for high birth rates to offset high death rates. It likewise suggests existence of characteristics described in the high stationary and early expanding stages of the demographic transition model.

Gross National Income Per Capital, 2001

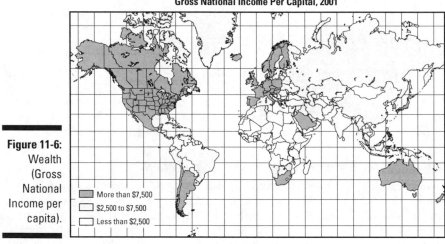

Figure 11-6: Wealth (Gross National Income per capita).

More than $7,500
$2,500 to $7,500
Less than $2,500

Percent of Population Under 15 Years of Age, 2001

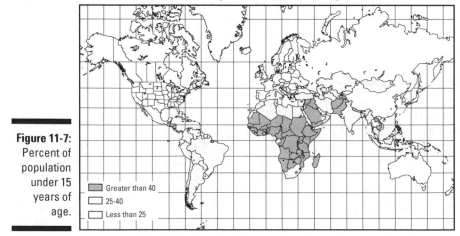

Figure 11-7: Percent of population under 15 years of age.

Greater than 40
25-40
Less than 25

Population pyramids

Population pyramids provide a graphic means of depicting and comparing the populations of different countries. In the diagrams below, the vertical axis shows age groups, while the horizontal axis indicates the percent of a country's population that is in each of those groups. Most countries have more young people than old people, so the graph typically has a wide base that tapers upward — rather like the shape of a pyramid. But the width of the base may vary substantially. Pyramids of developing countries, which typically have high rates of natural increase and therefore large percentages of their populations in the younger age groups, tend to have wide bases. In contrast, the pyramids of developed countries, with their low rates of natural increase, tend to have narrower bases.

Population pyramids are particularly useful for contrasting *dependents* and *non-dependents*. These are terms demographers use to characterize people in the under-15 and 65-and-over age groups (*dependents*) and the middle aged people *(non-dependents)* on whom they must generally rely for their sustenance. Typically, a low percentage of dependents is desirable because it means a high percentage of non-dependents is available to see to the needs of the young and the old. And in fact, that tends to be the case in well-off countries. In developing countries, in contrast, a high percentage of dependents relative to non-dependents is the norm. The population pyramids above make these differences very apparent.

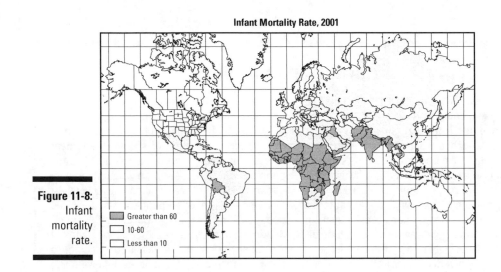

Infant Mortality Rate, 2001

Greater than 60

10-60

Less than 10

Figure 11-8:
Infant
mortality
rate.

Considering "Overpopulation"

Given the billions who live in poverty, poor health, and crowded conditions, people sometimes suggest that Earth is overpopulated. This controversial subject is made all the more problematical because it has proven difficult, if not impossible, to define. The clear implication of "overpopulation" is "too many people." But how many is too many, and is that number the same everywhere, or is it dependent on local conditions?

Reputable demographers agree there is no "magic number" of people or of people per square mile beyond which a country or region is overpopulated. Being statistically inclined, however, they do look to numerical data and analyses to gain perspective. Perhaps their most intriguing concept is *carrying capacity* — the number of people that a country or region can sustain at an acceptable level of well being given its prevailing technology.

As you will see in the "Regarding overpopulation and carrying capacity" sidebar, one can argue that technologically advanced societies have higher carrying capacities than developing nations that lack similar expertise. But differences in culture, life experience, and personal preference have generally rendered inconclusive mathematical attempts to precisely determine carrying capacities. At best, and in response to the questions above, demographers' numerical exercises suggest it is impossible to determine how many is too many, and that thresholds — if and as near as they can be determined — do vary from place to place depending on local conditions.

Lack of conclusive definition and indicators of overpopulation — statistical and otherwise — has not prevented people from taking sides on this issue. Some argue passionately that Earth or parts thereof are overpopulated and espouse policies to rectify the perceived problem. Others argue just as passionately that overpopulation does not exist and espouse policies aimed at relieving malnourishment, poor health, and its other would-be symptoms. Proponents of these viewpoints occasionally and respectively are referred to as *neo-Malthusians* and *cornucopians*.

Neo-Malthusians

If you hail from the camp of neo-Malthusians, your camp leader is Thomas R. Malthus (1766-1834), an English political economist and theologian, who believed population increase is a prelude to disaster. In his famous "Essay on the Principle of Population" (1798) he stated that "population increases in a geometric ratio, while the means of subsistence increases in an arithmetic ratio." In other words, human population is increasing and at a rate much faster than food supply.

Regarding overpopulation and carrying capacity

Below are selected demographic data for two honest-to-goodness countries. If I showed only the first two rows of data and then asked which country could be considered overpopulated, I suspect most people would say Country Y. After all, it has about 50 percent more people than Country X, and has a population density (people per square mile) that is more than 47-times greater than its counterpart's. But when you throw in the last three rows of data, a different picture emerges.

	Country X	Country Y
Population (millions)	10	16
People per Square Mile	21	1,018
Female Life Expectancy	41	81
Infant Mortality (per 1,000)	123	5
Per Capita GNI ($US)	740	24,410

Women live twice as long in Country Y, and infant mortality is nearly 25-times *less* prevalent. Also, per capita GNI data suggest personal income is astronomically higher in Country Y. The implication, therefore, is that citizens of Country Y are generally well off while those of Country X are not. This leads to consideration of carrying capacity — the number of people a country can sustain (carry) at an acceptable level of well-being. Review of the data suggest Country Y has a high carrying capacity. Though densely populated, it clearly has the capacity to carry its relatively large population at a high level of well-being. The opposite might be said of Country X. Even though the country is lightly populated, it appears not to have sufficient resources to sustain its population at a high level of well-being. Therefore, one may argue that Country X has exceeded its carrying capacity. What is certain, however, is that overpopulation cannot be determined simply on the basis of a country's population or its density. By the way, Country X is Niger, and Country Y is The Netherlands.

During Malthus's time, agricultural technology offered limited prospects for improved harvests. Thus, the only real option, as Malthus saw it, was for humans to have fewer babies. Artificial means of birth control were available, but Malthus opposed them on theological grounds. He preached restraint, but conceded that human passions were not likely to be held in check by "Just say no." Thus, he saw no solution save the grim reaper. Population would continue to increase faster than food supply until, ultimately, large-scale famine and starvation rectified the imbalance.

The world has changed a lot since Malthus. Human numbers have exceeded his wildest dreams, but so has agricultural productivity. Also, in Malthus's time, transportation technology was such that food had to be produced fairly close to consumers. Nowadays, however, food travels hundreds — even thousands — of miles to get to your supermarket. Thus, a local crop failure or poor harvest need not have the devastating effects of yesteryear because food can be brought in from somewhere else.

Today few reputable scholars espouse the literal word of Malthus. But lots of neo-Malthusians believe the old bloke basically got it right. Namely, many countries suffer from too many people (see Figure 11-9). And while technology may someday improve the average welfare, population reduction is the most effective and reliable way to achieve a better balance between people and resources.

Figure 11-9: A photo of a slum in a Third World country. The low quality of life in places such as this is central to neo-Malthusian arguments to limit population.

Cornucopians

This term *cornucopia* recalls "the horn of plenty," that curly-cued overflowing basket of food one tends to associate with Thanksgiving decorations. If you camp with *cornucopians,* the viewpoint is that the world has a food supply problem, not a people supply problem. And in their view, the solutions are not futuristic. Rather, means are available now to greatly increase global food supply and therefore improve carrying capacity throughout much of the world. They include:

- **Greater use of green revolution know-how.** *Green revolution* refers to a number of agricultural innovations designed to increase food production in developing countries. Chief among them are varieties of rice and wheat that have been genetically engineered to increase their yields as well as their resistance to crop diseases. Increasing access to these relatively inexpensive strains could greatly help developing countries to increase their carrying capacities.

- **Improved grain storage.** In several countries a significant portion of grain harvests are lost to vermin due to poor storage. Modest expenditures on secure storage could substantially increase available food.

- **Improved transportation.** Modest investment in the most basic forms of infrastructure could greatly enhance food supply and well-being. Roads in parts of many developing countries are little more than dirt tracks that may be nearly impassable during a rainy season or other time of year. This diminishes access to markets and to goods that might improve agriculture. Modest investment in road-building could lead to a much improved picture.

- **Greater distribution of food surpluses.** Some developed countries have massive food surpluses that are merely stored, as well as policies that pay farmers not to grow crops (in order to maintain decent price levels). Cornucopians regard these as ethically unconscionable and readily available sources of food.

Applied Geography: Census-taking from above

In developing countries census-taking is sometimes inhibited by the remoteness of villages and the reluctance of their inhabitants to be enumerated. Intent on conducting the best possible head count, some nations have successfully overcome these impediments through careful use of two geographic techniques: spatial sampling and aerial photography. Specifically, inhabitants in a number of accessible and representative villages throughout the country are surveyed (spatial sampling) with special emphasis on determining the average number of people per house or hut. Afterwards, aircraft fly over remote villages and photograph them. Photo interpreters then examine the pictures, taking special care to identify and count the houses and huts. That number is then multiplied by the average number of people per household, the result being the estimated population of the photographed areas. Geographic variation in house-types and social structure may complicate matters. But adequate sampling coupled with skillful photo interpretation may result in a reasonably accurate census.

Cornucopians, in short, believe that a number of means are available that can significantly improve global carrying capacity. What is lacking, in their view, is the will to do the right thing.

Figure 11-10.
A photo of a high-yield rice paddy. High-yield rice is one reason Cornuco-pians are optimistic about sustaining the world's people at an acceptable level of well-being.

Chapter 12

Shift Happens: Migration

. . . .

In This Chapter

▶ Studying early human dispersal

▶ Deciding where to go

▶ Marking migration magnets

▶ Marketing and mental mapping

. . . .

New Britain and New Holland are towns in Pennsylvania. New Prague and New Ulm are towns in Minnesota. And who could forget New Lisbon, Wisconsin; New Leipzig, North Dakota; and New Hamburg, New York? One can't help but wonder what was going on in Old Prague, Old Lisbon, and Old Leipzig that caused people to up and embark on the journey of their lives. Goodness knows, shift happens. And thank goodness it does, because geography would be pretty dull without it.

At issue in this chapter is migration, which is key to understanding

✔ The distribution of people at the global, regional, and local scales

✔ Differences in population growth

✔ Patterns of ethnicity and culture

✔ Environmental issues related to population growth

Migration is travel that involves a change in residential location. Together with birth rates and death rates (described in Chapter 11) it's a central component of population geography. More specifically, migration is fundamental to understanding population shifts present and past, including ones that ran their courses long before the dawn of recorded history.

Populating the Planet

Perhaps the greatest migration story of them all involves the populating of the planet. Scientific evidence suggests humans originated in East Africa. The Bible talks of the Garden of Eden, somewhere in Mesopotamia. Either way, the basic argument is that *homo sapiens* began by occupying one very small part of planet Earth.

That didn't last. When Columbus reached the New World, he discovered that other humans had gotten there long before he did. Throughout the Age of Discovery, other explorers also found that, time and again, other humans had beaten them to their newly found lands. Throughout the Americas, the Pacific Islands, the far Northlands, Eurasia, Africa, Australia, and New Zealand, you name it. Humans were just about everywhere except Antarctica. How had they done it? That is, assuming humans originated in a single region (and no credible evidence has been found to the contrary), how had they managed by 1492 to assume a near-global distribution? The answer is land bridges and ocean voyages.

Bridging the oceans

Once upon a time it was possible for humans to walk between certain continents and other land bodies that are today separated by straits and shallow seas. The key word in that sentence is walk — and on dry land, too. That was made possible by something rather peculiar that happened during the last ice age.

Earth has experienced several ice ages during the past million years. We're not certain what caused them, but climates then were definitely cooler on average than they are today. The most recent ice age began about 120,000 years ago and ended about 10,000 years ago. During that period, humongous amounts of seawater evaporated, condensed in the atmosphere, fell to Earth as snow, and compacted to form glaciers instead of returning to the sea as runoff. Thus, as the glaciers grew, sea level dropped. The exact extent of the decline is unknown, but for several thousands of years sea level was as much as 475 to 500 feet lower than today.

Thus, a world map at the height of the ice age would have looked a lot different from today's (see Figure 12-1). Substantial areas that had been ocean bottom became dry land, so the continents and other land bodies grew while the oceans shrank. Most importantly, several land bodies that had been separated by water became connected by *land bridges* — dry land in places that had been straits or shallow seas.

Figure 12-1:
The world at the height of the last ice age. The space between the bolded lines depicts what used to be solid land.

The land bridges lasted for thousands of years. Formerly ocean bottom, they became grasslands and woodlands that provided habitat for animals. Most importantly, of course, the land bridges provided firmament that allowed humans (over many generations) to migrate and occupy lands that had been unknown or out of reach. For example, people could have walked from present-day France to Ireland (see Figure 12-1). Similarly, of course, the ancestors of Native Americans migrated over dry land from Siberia to Alaska. Other connections are also evident on Figure 12-1, though some are contro-versial. We're not certain, for example, whether the Strait of Gibraltar was ever a land bridge and opinion varies as to the location of ice age coastlines in the Indonesian and Philippine archipelagos.

Eventually, the last ice age came to an end. Climates warmed, and as glacial ice receded their melt waters flowed to the oceans, which rose and inundated the land bridges. Thus, continents and land bodies became disconnected — just as they had been before the ice age started. But something significant had happened between the appearance and disappearance of the land bridges. Humans had migrated across them, giving rise to native populations that, many generations later, would greet European and other explorers.

Voyaging afar

Sometimes people were separated from their nearest neighbors by thousands of miles of ocean far too deep (14,000 feet or so) for any land bridge to explain their arrival. Therefore, ancient voyages of substantial scale must have occurred. Relatively recent research suggests the requisite navigational skills were based on knowledge and application of star positions, bird behavior, cloud types, ocean swells, and wave refraction. For example, when Captain James Cook reached the Hawaiian Islands in 1778, he found natives of Polynesian ancestry. Thousands of miles of ocean separated these people from any neighbors, proving that voyages had occurred. Other islands were found to have similar populations, and not just in the Pacific. Thus, the Merino people of Madagascar, off the southeast African coast, speak a Polynesian language.

Making colonial connections

The Age of Discovery foreshadowed an era of expansionism and colonial acquisition on the part of several European powers that would have major consequences for population geography. Specifically, exploration led to knowledge of and interest in distant lands that were perceived to have strategic or economic value — or both. These interests led to claims on territories that became colonies.

Establishment of colonies led to creation of *migration fields,* which are countries or regions that generate (such as the European powers) or receive (such as Australia, New Zealand, and parts of the Western Hemisphere) major migration flows. These migrations, in turn, led to the creation of regionally distinctive population characteristics that endure to this day. Thus, for example, what were once sparsely populated "native" lands in Australia, New Zealand, and parts of the Western Hemisphere now bear the unmistakable imprint of European settlement. In addition to free men and women, large numbers of indentured people — including non-Europeans — also relocated. Many Indians (by which I mean South Asians instead of Native Americans), for example, migrated within the British Empire and added significantly to population geography as far away as the Caribbean. The effects of all migration on native populations varied from elimination, to relegation onto reservations, and absorption or intermarriage.

Reciprocal flows of goods and people between European countries and their colonial possessions became commonplace. Low-cost raw materials were sent from colonies to colonizer, where they were made into higher-cost manufactured items and sent back to the colonies for sale. Thus began business and trade relationships that persist to the present, though not necessarily in the same, one-sided manner.

Likewise, in matters of migration, disproportionately strong migration fields exist among the countries of former empires. Thus, a visit to virtually any large city in England reveals ethnic neighborhoods dominated by West Indians, Pakistanis, and other ex-colonials. Similarly, a review of immigration in Canada reveals a migration field in which the countries of the former British Empire are disproportionately evident.

Forcing involuntary migration

Migration is not always a matter of personal choice. Sometimes it's involuntary — forced upon certain populations. Examples include the expulsion of Jews from parts of Europe, Native Americans from their homelands, and, more recently, various peoples from their homelands in the former Yugoslavia. But undoubtedly the most terrible case of them all, and the worst chapter of colonial history, is the Trans-Atlantic slave trade.

Beginning in the 1400s and continuing for nearly four centuries, untold millions of Africans were kidnapped and forcibly shipped off to the Caribbean islands, South America, and North America, probably in that order in terms of numbers of enslaved people. The impact is, of course, clearly seen today in the demographics of receiving areas, and perhaps in Africa as well. Much of the raiding that fed the Atlantic trade occurred in the latitudinal *Middle Belt* that lies between the underside of West Africa and the Saharan fringe. Demographers of the African scene have long regarded this area as underpopulated, and several explain that observation as an enduring legacy of the slave trade.

Choosing to Migrate

In a majority of cases nowadays, people migrate because they choose to do so. The decision to migrate varies from simple to complex. Sometimes the key element is a *push factor*. This is a characteristic of a region that causes dissatisfaction among residents and encourages them to emigrate. The most common push factors are war, political unrest, famine, persecution, dislike of the physical environment, and economic hardship. Push factors have been responsible for some of the greatest migrations in history. Emigration from Ireland due to the potato famine is a case in point. Another — and perhaps the greatest migration in recorded history — is the relocation of Muslims and Hindus within the Indian subcontinent after its partition into India and Pakistan.

 In other instances the decisive element in the decision to migrate is a *pull factor*. This is a characteristic of an area that exerts an attractive force that draws people from other regions. Being the opposite of push factors, pull factors include peace and harmony, lack of persecution, a pleasant environment, and economic opportunity. Many times, of course, both push factors and pull factors play a role in decision-making.

While push and pull factors encourage migration, potential *barriers to migration* have the opposite effect. These deterrents may include

✔ **Physical barriers** such as oceans, mountain, and deserts

✔ **Economic barriers** such as the costs of migration and of establishing a new home

✔ **Cultural barriers** that involve the sobering prospect of leaving a familiar religious, linguistic, and relational environment for an unfamiliar one

✔ **Political barriers** that may include policies of one's own country that discourage emigration, as well as those of potential receiving countries that discourage immigrants of one sort or another.

Numerous countries offer numerous examples of these and other migration concepts. Possibly none, however, surpass the United States.

Coming to America

The United States is often and justly referred to as a nation of immigrants. Prior to 1875, anybody from any foreign land could legally and freely enter the country and become a resident. Thereafter, Congress began passing a series of laws (which exemplify political barriers) that restricted immigration on such criteria as morality, race (starting with the Chinese Exclusion Act of 1882), and national origin. Laws in recent years have been characterized by annual caps on the number of newcomers and a system of preferences for family members, skilled workers, and people from under-represented countries.

Review of sources of immigrants reveals a significant change in migration fields. Considering U.S. immigration history from 1820 (the beginning of record keeping) to 1998, Europe was the principal source area, accounting for seven of the top ten countries overall (see Table 12-1). In various ways these Europeans satisfied the cost of migration (an economic barrier) and made passage across the Atlantic Ocean (a physical barrier). Once in America, of course, they often found that their Old World culture traits were obstacles (cultural barriers) to integration into to emerging social fabric. But migrate they did, ultimately because negative conditions in the Old World (push factors) and the attractions of the New World (pull factors) proved more powerful than barriers to migration.

Germany has the distinction of being the number one donor nation of all-time, but Mexico has been closing the gap at a torrid pace. The fact that Mexico recently passed Italy for the number two slot is impressive enough. The fact that about 60 percent of all Mexicans who have ever immigrated to the U.S. have done so since 1980 is astonishing. While future Americans still come from Europe, many more now come from Latin America and Asia. Indeed, countries in the latter areas accounted for all of the top ten sources of immigrants in 1998.

Table 12-1	Top 10 Sources of U.S. Immigrants Historically and Recently	
1820–1998	*1999*
Country – Number	***Country – Number***
Germany – 7,156,257	Mexico – 131,575
Mexico – 5,819,966	China – 36,884
Italy – 5,431,454	India – 36,482
U.K. – 5,247,821	Philippines – 34,466
Ireland – 4,779,998	Dominican Republic – 20,387
Canada – 4,453,149	Vietnam – 17,649
Former USSR – 3,830,033	Cuba – 17,375
Austria – 1,842,772	Jamaica – 15,146
Hungary – 1,675,324	El Salvador – 14,590
Philippines – 1,460,421	Korea – 14,268

Source: U.S. Justice Dept., 1998 Statistical Yearbook of the I.N.S. (2000).

Just as source areas have changed, so, too, have the final destinations of immigrants. Cities have always attracted large numbers of them, either as points of arrival or potential employment. But in the first half of America's history, an abundance of rural land (due to displaced Native Americans) was also available to pioneer settlers.

In many instances, immigrant groups of like origin from different parts of Europe settled large contiguous tracts, basically transforming the frontier into an ethnic quilt. Often, of course, settlements thrived, populations grew, and towns arose, resulting in the New Pragues, New Lisbons, and New Leipzigs that dot America. In these rural (or formerly rural) areas, land tends to be owned rather than rented, and therefore it gets passed down over the generations. As a result, a persistent ethnic geography that dates from pioneer days remains over much of America.

By the second half of America's history, the frontier was largely gone and with it the opportunity for new immigrant groups to settle large rural tracts. Accordingly, immigration assumed an increasingly urban focus, and so it remains. In cities, people tend to rent their residences rather than own them. Thus, the propensity is for ethnic turnover to occur in certain neighborhoods with each new wave of immigrants. As a result, if you live in New Prague, New Lisbon, or New Leipzig, then the countries in the right-hand column on Table 12-1 may surprise you. On the other hand, if you live in New York, Chicago, or Los Angeles, then this information may be yesterday's news.

Channelized migration, which links geographically specific points of origin and destination, characterizes the growth of several urban immigrant enclaves. For example, a sizable and growing Dominican neighborhood exists in Upper Manhattan. Detailed examination reveals that it is far too simplistic to describe the migration as people from the Dominican Republic moving to New York. Rather, people from particular towns or regions of the Dominican Republic are settling in particular parts of the ethnic enclave in Manhattan. Thus, specific channels are found within the overall flow of immigrants from the Dominican Republic to the United States. Similar examples appear in other parts of the United States regarding this and other immigrant groups. Similar examples have also been documented in other countries all over the world.

Neither channelized migration nor migration in general is strictly international. Many countries provide cases of internal (domestic) migration whose characteristics and impacts are no less significant. Again, the United States provides excellent examples.

Migrating at home

Domestic migration involves residential relocation within a given country. While some moves may involve crossing the street, others may be cross-country or inter-regional. The latter are of particular interest to geography for two reasons. First, and as we shall shortly see, they may be symptomatic of waxing and waning economies of different areas. Second, they may have significant political ramifications. In the United States, Canada, Britain, and other democracies, the number of legislative representatives allocated to a state, province, or region is based on the number of residents. In the United States, for example, the population of each state determines the number of U.S. Representatives that its citizens elect to Congress. As state populations rise and fall because of migration and other factors, the allocation of delegates to Congress changes, and therefore so does the geography of political clout.

Once again, it is useful to focus on a single country to examine the causes and consequences of domestic migration. And once again we will consider the United States.

Relocating within America

Americans are a people on the move. They commute, shop, take their kids here and there, and go away on vacation. But they also move in the sense of changing their residential locations. In fact, U.S. government data suggest

"The brain drain"

Brain drain refers to the global tendency for highly educated and skilled citizens of developing countries to migrate to the more developed nations. This population shift saps (or drains) the developing world of intellectual resources as it adds them to the already more well-to-do countries. People the world over seek education that prepares them for modern highly skilled, good-paying jobs. The economies of developing countries typically do not, however, generate large-scale employment of this type. In contrast and by their very nature, the economies of developed countries do. Indeed, in some developed nations the number of such jobs exceeds the indigenous supply of qualified applicants. One result is immigration laws in developed countries that give preference to highly trained foreigners who are prepared to fill the vacancies. In the United States, for example, the 1990 Immigration Act set aside 140,000 slots annually for foreigners with valuable skills. The years since have witnessed pressure to increase that number. The effect of these and similar policies is to increase the brain drain. While this policy clearly benefits the receiving developed countries, it also exemplifies the saying that "the rich get richer, and the poor get poorer."

that about 20 percent of Americans move each year, and that about 3 percent change their state of residence each year. The latter figure may not seem like a lot. But given some 285,000,000 Americans, that means about 8,000,000 people leave one state and move to another each year

Principally because of this domestic (that is, interstate) migration, the populations of states and regions of the United States are growing at different rates (see Figure 12-2). Nevada, for example, grew by an astounding 66 percent during the 1990s, while the population of North Dakota grew by only 0.5 percent. Overall, we can pinpoint discernible regions of growth. The highest rates of increase are occurring in the southern tier of states (the *Sunbelt*) plus the Northwest. The lowest rates of growth are occurring in the Northeast and Upper Midwest (*the Snowbelt*).

As implied above, these differences have little to do with natural increase and are only modestly related to immigration from foreign lands. Instead, the patterns in Figure 12-2 are largely a function of domestic migration in general and complementary migration fields in particular. That is, people have generally relocated from the low-growth areas to the high-growth areas.

This relationship has been in high gear for at least the past 30 years. Reasons for it include

- ✔ A surge in retirees migrating from the northern Snowbelt (a push factor) to the warmer climes of the Sunbelt (a pull factor)

- ✔ A rise in employment opportunities (pull factors) in the southern tier (and the southeast in particular) as businesses moved in to take advantage of the region's comparatively low wages, low taxes, and low energy costs (which also attract of their own accords good numbers of immigrants)

✔ A related rise in unemployment (a push factor) in the Northeast and Upper Midwest (due to the closure of numerous heavy industrial plants in the traditional manufacturing belt) and the desire and willingness of the jobless to relocate where jobs are available

✔ The attractive outdoor amenities and recreational potential (pull factors) associated with the climate, beaches, landscapes, and open spaces of parts of the South and West

Normally, one thinks of growth as a good thing. But some high population growth areas are now victims of their success at attracting migrants. Virtually every one of the 25 fastest growing cities in the U.S. are in the Sunbelt or West, where the imagery of open space and the great outdoors has met with traffic jam and urban sprawl (see Figure 12-3.) Other areas have witnessed a growing disparity between the need for water and the supply of it. Primary examples are California, Arizona, and Florida, where thirsty cities and even thirstier agriculture are increasingly pitted against each other.

Percent Population Growth By State, 1990-2000

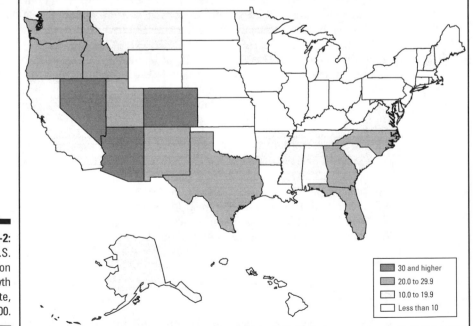

■	30 and higher
▨	20.0 to 29.9
□	10.0 to 19.9
□	Less than 10

Figure 12-2:
U.S. population growth by state, 1990–2000.

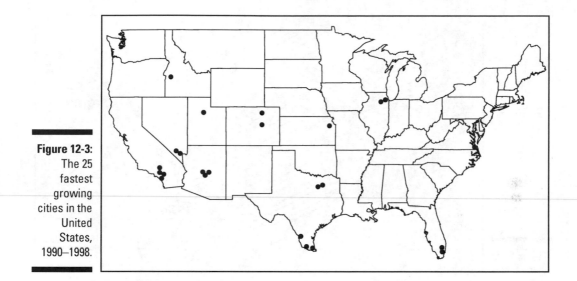

Figure 12-3:
The 25
fastest
growing
cities in the
United
States,
1990–1998.

Giving a Good Impression

Decisions to move to or vacation in particular destinations tend to be based largely on the knowledge and impressions people have about them. Different people have different attitudes toward different places. Some locales may impress you as desirable places to live, some may be undesirable, and some may make no impression one way or the other. Sometimes these attitudes are based on personal experience and verifiable facts (objective reality), and sometimes they are based on what people have heard or read or imagine to be the case (subjective reality). When you think about it, don't you have strong positive or negative feelings about some places, including ones you have never visited?

Playing the mental game

Mental maps are tools that geographers use to display and analyze the impressions (subjective realities) that people have about different locations. The example provided in Figure 12-4 shows a mental map of the United States that reflects attitudes of college students who live in New York City. Now, before somebody gets upset and writes a nasty letter to the publisher or me, remember that this map reflects attitudes that some people have toward places that perhaps they have never visited and about which they may have no direct personal knowledge. Thus, what they think they know about a given state or region may have no basis in reality. Moreover, just as some New Yorkers have negatives images about certain parts of the country, it's also true that people in other parts of the country have very negative attitudes toward New York.

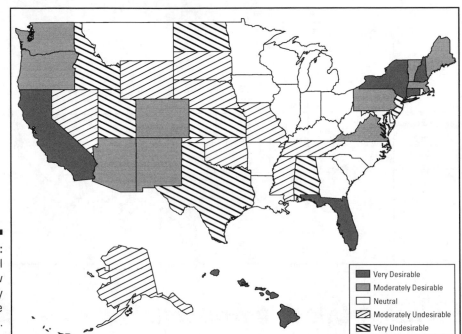

Figure 12-4:
A mental
map of New
York City
college
students.

Looking at the mental in Figure 12-4 may cause you to ask, "What was going through the minds of those students?" In fact, like most mental maps, this one reveals a handful of factors that repeatedly determine place desirability or undesirability. They include the following:

✔ **Home:** Dorothy said it best: "There's no place like home." No matter where you conduct a mental map exercise, you find that most people like where they are. Thus, New Yorkers like New York, Londoners like London, and so forth. Home, after all, is known and predictable and usually doesn't hold the nasty surprises that a move might.

✔ **Dislikable neighbor:** A dislikable neighbor often borders one's desirable home. Thus, many people in New York City have very negative perceptions of New Jersey. What this factor says about the human condition is open to debate.

✔ **Physical environment:** Some people perceive some states to be more environmentally attractive than others. Most college students in New York City, for example, generally have positive views of states that they perceive as having warm beaches, hilly and/or mountainous topography, and forests or related greenery. On the other hand, places that are cold, flat, and desert-covered are worthy of avoidance.

✔ **Socio-cultural environment:** People come in different races, speak different languages, adhere to different religions, have different sexual orientations, and so forth. Thus, when people assess the desirability of a particular destination, they often ask themselves, "Will the people who live there accept me, or will I find it difficult to fit in?"

✔ **Job prospects:** If you give a mental map exercise to people who are retired or about to be, then job prospects typically don't mean diddly. But because most college students are pondering employment after graduation, they are usually attracted to states that are viewed as offering good job prospects.

Getting an image adjustment

Mental maps can be a lot of fun to play around with, but they may strike you as rather trivial. In fact, they have a very serious side as predictors of potential migration. Most places seek to attract entrepreneurs and businesses that can create jobs and generate tax revenues. Mental maps can sometimes reveal that a particular place has what may be called an "image problem" in certain parts of the world, or is simply not seen as a great place to do business.

This perception may encourage officials in that place to create ad campaigns or engage in other marketing measures designed to promote a positive image. If you routinely watch TV or listen to the radio, then sooner or later you are bound to see or hear a commercial that encourages people and businesses to move to a particular place. The reason is simple. Officials who live in that "particular place" know that shift happens. They also know that getting it to happen in their direction can add to employment, tax revenues and political clout.

Putting your best image forward

I cannot end this chapter without a word or two about tourism — a migration of sorts that has much to do with geographical impressions and subjectivity. Tourism is a multi-billion dollar industry that is likely to grow as more countries develop and more people have more disposable income. Seemingly every country and locale wants a piece of the multi-billion dollar travel and tourism pie. They won't get a sliver, however, unless tourists perceive them as a desirable place to visit.

Enter advertising, which is a major means by which tourists find out about and assess destinations that are anxious to attract them and assist in the disposal of their disposable income. In the world of tourism advertising, objective geography tends to give way to fantasy geography. It's not that tourism ads lie, but they definitely do put a certain spin on reality. Thus, I have yet to see a travel ad that shows a rainy day (even for Ireland, which isn't called the Emerald Isle for nothing). No matter the destination that is being touted, the landscape is always gorgeous, the locals are always smiling (and often dressed in traditional garb that nobody really wears anymore), and the tourists are healthy and fit. And why not? The goal is creation of a positive image that potential tourists will find attractive. Will tourists visit places they perceive as unattractive? Would you?

Applied Geography: Marketing Venezuela

Some time ago the Venezuela National Tourist Office embarked on an advertising campaign to promote tourism in Caracas, the capital city, and Margarita, a resort island. The ads ended with the words "Venezuela, the country in the Caribbean." One thing the ads did not include was a map. What it would show, of course, is that Venezuela is in South America while Margarita is indeed in the Caribbean. But "country *in* the Caribbean"? Venezuela is certainly *by* the Caribbean, or *borders* the Caribbean, or *lies alongside* the Caribbean. But *in* the Caribbean? Close, but no cigar.

Why would a country in South America profess to be in the Caribbean? The answer is that potential tourists from the United States have a much more positive image of "Caribbean" than "South America." The former evokes images of sun, surf, palm trees, and piña coladas. The latter (at least as some are concerned) evokes images of drug cartels, slums, leftist guerrillas, and cholera epidemics. So why not call yourself "the country in the Caribbean" as long as there is an ounce of truth in it?

If this sounds a bit duplicitous, you're right. One application of geography is to selectively filter the information people obtain about a place so as to manipulate perceptions of potential tourists. Another application, however, is to help people see through the smoke screen and perceive the whole truth, not just an ounce of it.

Chapter 13

Culture: The Spice of Life and Place

- -

In This Chapter

▶ Creating cultural diversity

▶ Spreading culture, stopping culture

▶ Considering religion and language

▶ Heading toward a single global culture?

- -

*H*ave you ever stepped on somebody while he was praying? I have. And I would have loved to crawl under a rock, except there weren't any.

The incident happened in Dakar, Senegal. It was a Friday, and I was out walking around the streets of that West African capital, playing tourist. The Senegalese are overwhelmingly Islamic. Devout Muslims pray several times daily, and Friday prayers have special importance. So there I was, just your basic Presbyterian walking along Dakar's version of Madison Avenue, when the call to midday prayer rang out. This is surely an exaggeration, but it seemed that in seconds traffic stopped and people poured out of buildings and filled the street. Many of them had prayer rugs. Others carried substitutes, such as newspapers or pieces of cardboard. Soon the street was covered by people performing their solemn religious duty.

For some reason, I felt it advisable to get to the other side of the street, and in the process I stepped on somebody. I apologized profusely but the offended party remained completely focused on his prayer.

Everybody possesses *culture,* a learned pattern of behavior that characterizes a group of people. In many respects, your culture defines your essence; and yet you tend not to think about it until perhaps, you find yourself in a geographic setting where your culture either doesn't work, or marks you as being different from others. Suffice it to say, on that Friday noon I was suddenly and acutely aware of my culture.

Cultural geography is the field that seeks to describe and analyze the distribution of culture over Earth's surface, and is the subject of this chapter. But what can one say about the geographies of so many culture groups and their cultural traits in one chapter? The answer is "very little." Thus the emphasis of this chapter is on key concepts of cultural geography and how they affect the two major culture traits of religion and language.

Being Different 15,000 Times Over

Nobody knows exactly how many cultures exist on Earth today, but numbers like 15,000 tend to get thrown around. Whatever the true total, like birds, folk of a feather flock together. That is, people who are culturally similar tend to live in proximity to each other and in doing so, form *culture areas* — regions occupied by people who have something cultural in common. These may be quite large, like the Islamic culture region that extends from Dakar eastward across Northern Africa, through the Arabian Peninsula and Southwestern Asia. Culture areas may also be rather small, as in San Francisco's Chinatown, which occupies no more than a square mile or two.

In the act of practicing their culture, humans often transform the natural landscape into a *cultural landscape,* as when people convert a grassland to a farm. Because culture is diverse, so, too, are the world's cultural landscapes, which easily rival (and maybe surpass) purely natural landscapes in their richness and variety (Figure 13-1a and b). Culture therefore distinguishes people as well as physical regions. In a sense, culture is the spice of life as well as place, differentially "flavoring" people and the land they inhabit.

How did we get 15,000 cultures? Assuming human beings started out more or less the same way back when, then how did we end up so different? The answer is largely in three parts: the diversity of culture, effects of isolation, and adaptation to new surroundings.

Counting cultural diversity

Culture is extremely broad and complex, affording ample opportunity for people to be different from each other. Suppose you made a list of all the ways in which you are culturally different from people who live in Saudi Arabia, Thailand, and Tahiti. You could well end up with at least a couple of dozen items in each case.

Comparison of those lists would reveal *cultural universals,* which are categories of traits that all cultures share, but whose specific manifestations vary from one culture to the next. Various cultural universals are given in Table 13-1. Language is an example. You speak at least one language. So do people in Saudi Arabia, Thailand, and Tahiti. Indeed, every culture has one, so it can be considered a universal. All told, an estimated 6,500 languages are spoken today. Likewise, religion is a universal, and hundreds, if not thousands, of them can be found throughout the world. You can add to them dress, architecture, sport, and all of the other universals, each of which come in many specific varieties. If you count all of the ways different human groups have combined different manifestations of these universals, then you have the number of cultures on Earth. And apparently, that number is about 15,000.

Table 13-1		Some Cultural Universals		
Language	Religion	Dress	Technology	Architecture
Government	Cuisine	Economy	Dance	Gender roles
Grooming	Music	Sport	Law	Education
Medicine	Agriculture	Land tenure	Kinship system	

Isolating people

Isolation is another reason why we have so many cultures. Communication generally breeds cultural homogeneity. In other words, the more people that share information and ideas, the more alike they tend to be. Geographic isolation, in contrast, breeds differences. Take a large number of people who have the same culture, divide them into, say, four groups that are isolated and completely out of touch with each other, and over time they are likely to go their separate ways, culturally speaking. Basically, as humans migrated eons ago from their common ancestral homeland, to ultimately occupy the world, that is what happened.

Figure 13-1a and b: When peoples of different culture practice agriculture, they may transform the natural landscape into very different cultural landscapes.

Table 13-2 shows the conjugation of the verb "to sing" in four languages. Visual comparison of the columns shows different spellings but also remarkable similarities.

Table 13-2	How People "Sing" in Four Languages			
Verb Tense	*English*	*German*	*Dutch*	*Swedish*
Present	Sing	singen	zing	sjunga
Past	Sang	sangen	zong	sjong
Participle	Sung	gesungen	gezongen	sjungit

The explanation for the similarities in the spellings is that these modern languages have a common ancestral language that was spoken more than a thousand years ago in Central Europe.

But there came a time when members of this culture began to migrate, ultimately to occupy lands that would later become England, The Netherlands, Sweden, and Germany. And in the process, groups of significant size became isolated from each other more or less permanently. Patterns of speech were no longer shared. Pronunciations began to drift apart, which accounts for the differences in spellings. Migrants encountered "strange" plants, animals, and physical environmental conditions for which there were no words, so they invented new ones. Over time, therefore, the once-common language developed different dialects, which became different languages.

Adapting to new surroundings

Adaptation refers to cultural adjustments that occur after a people migrate to a region whose physical environment is different than where they used to live. A reference was made in the previous section to how a single language became several languages in part because of human adaptation to new environments. Here are examples of how adaptation has encouraged diversity in three other cultural universals:

✔ Traditional agricultural know-how may not work or is inefficient in the new setting. Immigrants borrow workable techniques from the local populace (if there is any), experiment with local plants for their food value, and otherwise adapt as best they can.

✔ Traditional modes of dress are either too light or too heavy for the new climatic venue. Adaptations may occur with respect to style, materials, and color.

✔ Traditional building materials may be absent in the new setting, resulting in adjustment in architecture. Also, climate may affect modification of housing with respect to thickness of walls, shapes of roofs, and openness to outside conditions.

In each case, adaptation that adds to the over all richness and variety of culture, increasing the ways in which people can be different.

Spreading the Word on Culture

Culture creation and modification are not things of the past. Thanks to *cultural diffusion,* which refers to the spread of culture, the geographies of particular traits as well as cultural complexes that characterize groups of people continue to develop and change.

For example, 50 years ago few Americans had heard of yogurt, tortillas, tofu, tandoori chicken, sushi, and couscous, let alone actually eaten them. All those foodstuffs existed back then, but their geographies were pretty much limited to their respective native areas — Asia Minor, Mexico, China, India, Japan, and North Africa.

Today it's different, of course. Chances are you have heard of most or all of those foodstuffs and maybe even eaten them — perhaps because you have traveled to their native lands, but more likely because those foodstuffs have spread here and are widely available in stores and restaurants. As a result, the geography of these foodstuffs, all of which are culture traits, has changed dramatically thanks to cultural diffusion.

The same is true of other traits. You, for example, speak one or more languages and may practice a certain religion. Chances are good, however, that none of these traits originated right where you live, but rather, like those foodstuffs, diffused from somewhere else. The manner of their diffusion may have varied, but probably has something in common with one or more of the three generally recognized modes of diffusion (see Figure 13-2): relocation, contagious expansion, and hierarchical.

Relocating one's culture

Relocation diffusion is synonymous with migration (Figure 13-2a). When people move, they take their "cultural baggage" with them. As a result, the geography of culture may change because migrants impart their particular cultural characteristics to an area where perhaps it was not previously present.

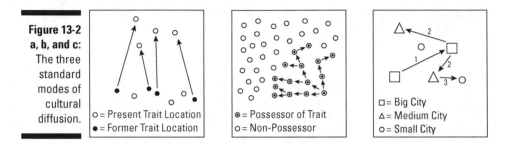

Figure 13-2 a, b, and c: The three standard modes of cultural diffusion.

Virtually every large American city, for example, has distinctive ethnic neighborhoods that exist because of the relocation diffusion of peoples from a foreign land. Small towns may also exhibit the same effect, and so, too, rather rural parts of America. Chances are good that an example or two are near you wherever you live. Indeed, *you* may be an example.

Coming down with culture

In **contagious expansion diffusion,** the geography of a trait expands because people who did not previously possess it, adopt it. Typically, this results from contact or direct exposure (hence, contagious) to the trait (Figure 13-2b). Thus, a farmer might "look over the fence" to see a neighbor growing some new kind of crop, and adopt it as well.

That is not a far-fetched scenario. Efforts to increase food production, for example, often are an exercise in cultural diffusion, as when an **agronomist** or crop scientist seeks to encourage local farmers to discontinue a traditional way of producing foodstuffs and do something different. But whether they live in Kenya or Kansas, farmers tend to stick with things that work and try something new only when the likelihood of success is high.

Demonstration is a proven way to promote diffusion of agricultural innovation. Neighboring farmers "look over the fence," to see what's happening in the demonstration area, and adopt it. Their farms, in turn, become objects of observation by other farmers who look over the fence, adopt what they see, and so forth.

OK, so you're probably not a farmer. But at some point in your life, you probably adopted a certain cultural item because of direct exposure to advertising, or because you saw somebody doing something or wearing something that you found attractive, or because of something a friend said or did. If so, then you have experienced contagious expansion diffusion firsthand.

Applied Geography: Looking Swiss

Drive into New Glarus, Wisconsin and you might get the impression that a little piece of Switzerland has been transplanted in the American Midwest. That's what the locals are hoping — and even more so that you will stop and spend some money. As in many other towns and venues in America, the people of New Glarus have taken to using their cultural heritage to promote tourism and the local economy.

In the 1840s immigrants from Canton Glarus, Switzerland settled the area, bringing their culture with them. The ensuing relocation diffusion of their German language, Catholic religion, dairying, and other attributes resulted in creation of a culture area in south-central Wisconsin that was distinct from neighboring lands peopled by immigrants from other European countries, as well as the local Native Americans. Other than dairying, however, little in the cultural landscape was particularly "Swiss-looking." That started to change a few decades ago when townspeople began erecting buildings in traditional Swiss styles and altering existing facades to render the same visual effect. The presumption was that if you turned the town into a "little Switzerland," then tourists would come and spend money. It worked like a charm.

You don't have to go to New Glarus to appreciate the concept, however. Lots of other towns and neighborhoods have done the same, albeit in all likelihood through a different cultural heritage. Perhaps you can think of an example or two in your own area.

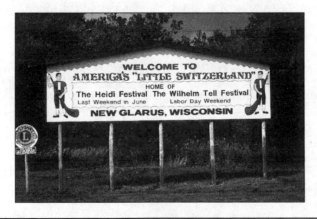

Doing what the big boys do

In *hierarchical diffusion,* a culture trait is born in a large city, becomes adopted by a portion of the populace, spreads to other large cities, and then "trickles down" to medium-sized cities, small cities, towns, and villages in that order (Figure 13-2c). Derogatory reference to this process is found in terms such as "country bumpkin" and "hicks from the sticks," which typically are applied to rural or small town residents who are allegedly "behind the times" with respect to cultural trends.

Nowadays, cultural "fads" in particular tend to diffuse hierarchically. This is especially true of new clothing styles, body modifications (hair styling, tattoos, piercing, and the lot), and slang. Inhabitants of large cities generally are more dissimilar and accepting of personal differences than their counterparts in small settlements where, perhaps stereotypically, everybody knows everybody else's business and pressure to conform is comparatively high. "Being different" by adopting a new fad that may seem outlandish to some is easier in big cities. If the fad is successful and catches on, then its visibility and acceptability are likely to increase, which in turn increases the likelihood that it will trickle down the hierarchy.

Calling a Halt: Barrier Effects

Barrier effects are things that stop or inhibit cultural diffusion. When culture traits spread, they typically do not "keep going and going and going" like that battery-powered bunny of TV commercial fame. Instead, traits tend to diffuse outward from their areas of origin, achieve a certain geographic breadth, encounter one or more barrier effects, and then stop spreading. Were there no barrier effects, then culture traits would, in fact, "keep going and going," resulting in a rather uniform global culture. Thanks to barrier effects, therefore, Earth's cultural geography is a mosaic of culture areas instead of a monochrome.

Barriers may be **absorbing** or **permeable,** respectively stopping completely the spread of culture or selectively accommodating the spread of some culture traits, but not others. For millennia, the Atlantic Ocean was an absorbing barrier that stopped the westward expansion of European culture. More recently, societal decision making in Saudi Arabia has been a **permeable barrier,** allowing into that country western technology related to oil drilling, but holding at arm's length other western cultural commodities such as bikinis and beer. As these examples indicate, barrier effects may originate from the physical or social environment.

Getting physical

Physical barriers are natural elements that now or in the past inhibit cultural diffusion. These have historically served to isolate people, either preventing or seriously limiting access to agents of culture change. The following sections cover the classic examples.

Oceans

Oceans were formidable barriers to cultural diffusion for millennia. People didn't know what lay across them, or how far away places were. Similarly, they did not possess the technology to accurately plot a course to a particular destination or to return home whether or not they had discovered anything. On top of that, and for the longest time, ships were fragile and at the whimsy of wind and storm. Thus, until modern shipbuilding and navigation came along, oceans tended to inhibit the spread of culture instead of promote it. To this day scattered Pacific Ocean Islands are homes to people whose cultures have been only modestly (if that) altered by contacts with the broader world. In these cases, the surrounding ocean continues to serve as a formidable physical barrier that has insulated the islands from forces of culture change.

Forests

Most of today's forests are mere remnants of their former selves. Five hundred years ago, nearly all of what is now the United States east of the Mississippi River was continuous forest — as was most of the Far West and Northwest. The same was true of virtually all of Western and Central Europe, as well as virtually all of humid Africa, Asia, Central and South America. And I don't mean the well-tended greenery you see today in many places. No sirree. I'm talking underbrush and thickets and dead limbs and all kinds of other stuff that limited visibility and mobility. It was ripe for disorientation and ambush. You could get lost in it.

And in a sense, that is what happened. Numerous peoples became separate, forest-dwelling societies whose woodsy surroundings provided isolation that contributed to development of distinctive cultures. Today numerous traditional societies inhabit regions of tropical rainforest, particularly in the Amazon Basin, but also in Central Africa and Southeastern Asia. Road building can be very difficult in these areas, and thus the forests, like the oceans in the case above, continue to isolate inhabitants from the outside world and promote cultural differences. In these environments, rivers — natural highways — have often served as avenues of diffusion.

Mountains

Rugged terrain, and particularly mountains, has historically tended to make communications difficult, and thereby encourage cultural diversity. For example, an estimated 700 languages are spoken on the island of New Guinea, which is about the size of Texas and Arkansas combined and has a population of perhaps 7 million. It makes no sense that so many languages coexist in such a relatively small space until you consider the topography. New Guinea has an extremely mountainous spine that has been eroded over the years into numerous steeply sided valleys that have no roads and few tracks between them. Add the dense tropical forest, and the results are hundreds of relatively isolated pockets of people that have, at least with respect to language, gone their own ways.

Also, the "thin air" that comes with the high altitudes of mountainous terrain has proven to be an impediment to diffusion of culture. For example, many facets of traditional Native American culture are alive and well in the Central Andes, where millions of people still speak Quechua, the language of the Incas. Although these peoples came under Spanish rule, the Spaniards themselves generally avoided settling in the high Andes because they found adaptation to the "thin air" to be extremely difficult. Accordingly, native culture in that area did not give way to imported culture.

Deserts

Outsiders generally were not adapted to desert conditions and therefore they found such regions inhospitable and avoided them. Accordingly, deserts have tended to isolate people and inhibit the spread of culture. For example, traditional culture groups continue to inhabit central desert areas of Africa, Australia, and Asia. The Bushmen of Namibia and aborigines of Australia are historic examples, although members of these groups have experienced significant change in recent decades. Nevertheless, the long persistence of their unique cultures testifies that deserts create a formidable physical barrier.

Tundra

Tundra, which you can read about in Chapter 10, refers to very high latitude environments dominated by short grasses. The climate is sub-freezing for much of the year. Native peoples adapted to these harsh circumstances over the years and developed distinctive cultures. Like deserts, however, outsiders generally are not well-adapted to tundra, and therefore have found such regions inhospitable and avoided them. Thus, tundra has served as a physical barrier. Specifically, it has tended to isolate traditional peoples in northernmost North America and Europe from sources of culture change and in doing so encourage a world of cultural differences.

Socializing effects

Social barriers are human institutions that inhibit the spread of culture. These can be as formidable as physical barriers, and sometimes more so. The following sections discuss four of the more prominent examples of social barriers.

Language

The simple inability to speak somebody else's language limits opportunity for cultural interaction and sharing. Though media and multilingualism are bringing people closer together, the continued existence of several thousand languages remains a powerful barrier to cultural diffusion.

Religion

Differences in religious beliefs may mark certain people as being "other" and nullify propensity to interact with them and adopt their cultural attributes. Religion may also manifest prohibitions (such as bikinis and beer) that deter exchange of materials and ideas.

Race and ethnicity

Many people have a deep "consciousness of kind." In some cases, that is code for racism and prejudice — a desire to be geographically separate from

The geography of Gullah

Gullah is a dialect spoken by African Americans of the Sea Islands, which adjoin the coasts of Georgia and South Carolina. Closely related to an ancestral language (or languages) that arrived with enslaved Africans, *Gullah* is spoken not by people who learned it to get in touch with their African roots, but instead by folks who never lost those roots in the first place.

The geography of *Gullah* is a classic example of physical barrier effects. After emancipation, most of the black population of the Sea Islands stayed there, and for decades the islands largely remained unconnected by bridges to the mainland. The result was limited physical inter-action with outsiders and mainstream culture. Under those circumstances Gullah continued as a viable language.

But times are changing. The Sea Islands, with their seaside settings, have become prized real estate for vacation homes, retirement communities, and resorts. Bridges now connect the mainland to many of these islands. As newcomers come and development physically transforms the islands, so, too, will they bring culture change that will threaten the continued existence of Gullah.

"them." In others, it may simply be a deep-seated preference for interaction with one's own kind. In any event, race and ethnicity tend to differentiate human groups — often very visually — and give rise to behaviors that limit interaction.

Historic events

Sometimes historic events, such as conflict and war, mark a people as "the enemy" and create wounds that refuse to heal. Intense dislike — if not raw hatred — of one group for another can be a powerful barrier to human contacts that would normally promote cultural transfer.

Getting Religion: How It Moves and Grows

Now let's see how the concepts mentioned up to this point come into play, beginning with religion. Looking at the geography of religion allows you to examine one of the more important culture traits and to make connections between cultural geography and contemporary matters. In Figure 13-3, a highly generalized map of the world's principal religions reveals culture areas that vary greatly in size. Christianity and Islam, for example, exhibit multi-continental expanses. Judaism, in contrast, is dominant in Israel, a comparatively small culture area, and in even smaller scattered urban enclaves mainly in Europe.

Putting diffusion to work

With the exceptions of the traditional (or what some people might call "tribal") religions, virtually all of the distributions shown in Figure 13-3 are results of cultural diffusion. In that regard, Buddhism and Christianity are interesting in that their present distributions have little in common geographically with where they originated — Israel and the West Bank in the case of Christianity, and India in the case of Buddhism. The Jewish population of Israel largely consists of recently (post-World War II) relocated individuals and their descendants, while the other Jewish enclaves worldwide are to some extent latter day expressions of the ancient *Diaspora* (the dispersion of Jews after the Babylonian exile). In Africa, Europe and Southwest Asia, the Islamic realm is largely the result of relocation diffusion (migration and conquest) of people from Arabia coupled with conversion of peoples with whom they came into contact (contagious expansion diffusion). Another interesting geographic aspect of Islam is its dominance in Indonesia (which contains more Muslims than any other country on Earth), Malaysia, and the southern Philippines, which are products of ages-old trans-Indian Ocean trade between Middle Eastern and Southeast Asian lands (contagious expansion diffusion).

Getting effects into action

To a certain degree, barrier effects are also evident on the map. The Himalayan Mountains, a formidable physical barrier, mark the boundary between parts of the Hindu and Buddhist realms. In West Africa, the interface between the Islamic and "traditional religion" realms coincide with the tropical forest fringe. A similar effect is seen in the Amazon, where tropical forests have isolated practitioners of traditional religions from agents of culture change.

Creating local character

In exercising their faiths, humans often impart religious character to the lands they occupy. The nature of these impacts vary from creating cultural landscapes to making psychological attachments to place, and from performing acts of solemnity to committing acts of bloodshed. Here are four ways adherents imbue locations with religious characteristics.

Places of worship

Practitioners of some religions build houses of worship that are magnificent works of architecture and, quite literally, outstanding components of the cultural landscape. Until fairly recent times, the spires of churches, towers of mosques, and domes of temples, dominated skylines. Indeed, in many smaller towns and some cities worldwide, they still do.

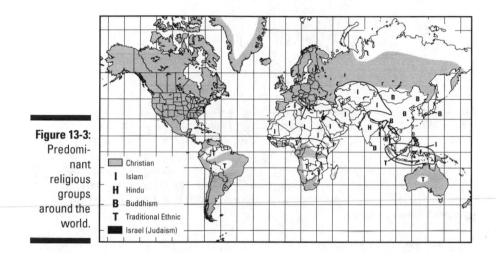

Figure 13-3: Predominant religious groups around the world.

Legend:
- Christian
- **I** Islam
- **H** Hindu
- **B** Buddhism
- **T** Traditional Ethnic
- Israel (Judaism)

While the essence of place is usually captured in things that are seen, it may also be echoed in things that are heard. Thus, some cultural geographers speak of "the audible landscape." That may sound a bit cryptic, but if you have ever heard church bells reverberate through a valley, or heard the Islamic call to prayer ring out from aminaret, then you have experienced the power of sound as a cultural geographic characteristic.

Sacred sites

Most religions recognize sites that have special significance in the minds of believers. These may serve as destinations of pilgrimage (such as Mecca and Lourdes), places of great historical religious significance (the Wailing Wall), or places where a solemn ritual is to be performed (Benares or the Via Dolorosa). The bond between religion and sacred site may be so strong as to create an uncompromising sense of proprietorship and right to rule in the sacred area, perhaps coupled by exclusion of non-believers.

Friction and flash points

Religions define behaviors that are pleasing and displeasing to God. Unfortunately, behavior (such as the proper way to call and worship God) that is practiced by members of some faiths may be displeasing or patently offensive to members of other faiths, leading to friction, if not outright bloodshed. It should come as no surprise, therefore, to observe contemporary conflicts that coincide geographically with the overlap or interface of different religious groups. Examples include:

- ✔ Northern Ireland (Protestant — Roman Catholic)

- ✔ Kashmir (Hinduism — Islam)

- ✔ Southern Philippines (Christianity — Islam)

✔ Israel (Islam — Judaism)

✔ Former Yugoslavia (Christianity — Islam)

Forbidden and favored foodstuffs

Religious dictates have changed the agricultural landscape of many areas by discouraging production of some foodstuffs and encouraged others. These have had a significant impact on the geographies of agriculture, animal husbandry, and cuisine. Among the better known prohibitions are the Judaic and Islamic proscriptions against pork (pigs being viewed as unclean), the Hindu proscription against beef (the killing of cows being forbidden), and the Islamic proscription against consumption of alcoholic beverages. In some cases, in contrast, religious favor has encouraged diffusion of foodstuffs. Here are two examples:

✔ **Tumeric:** This is a rather tasteless spice, grown throughout much of Southeastern Asia, whose principal purpose is to add a golden color to other foodstuffs. Its diffusion has been linked to the spread of Buddhism, in which gold is a symbol of enlightenment (hence the classic gold-domed Buddhist temples). The spice's purpose, therefore, is to bestow a desired aura on food.

✔ **Grapes:** The diffusion of Christianity, especially Roman Catholicism, in many parts of the world was complemented by the spread of *viniculture* (grape growing) to meet sacramental needs. In time, of course, grape production expanded to meet consumption needs of a more pedestrian sort. But religious diffusion provided the impetus for agricultural diffusion. Interestingly, the religious connection is particularly visible in Germany in the guise of the famous "Blue Nun" label, and the popular white wine *liebfraumilch,* meaning literally "milk of the blessed Lady," Mary.

Getting in a Word about Language

Language is arguably the most important of the cultural universals that were identified earlier. This is not to question the significance of religion or other traits; but language is essential to communicating and sharing many aspects of culture. The standard first step in analyzing the geography of languages is to produce a map of them. Unfortunately, on a page of this size, I can't possibly give you a map that shows the geography of the estimated 6,500 languages that are spoken today, nor even a map showing the dominant language in different parts of the world. Simply too many are out there. But I can, at least, show the geography of English usage worldwide (Figure 13-4).

What Figure 13-4 shows are large-scale English language culture areas around the world. In Britain, Australia, New Zealand, Canada, and the United States, English is spoken by the overwhelming majority of the population. In other countries, English is spoken only by a minority, even though it may be an "official language." The "big picture" map aside, consideration of language — like religion — affords opportunity to observe and apply diverse concepts of cultural geography.

Diffusing languages

The map of English is in large measure a product of cultural diffusion from Britain through its former colonies. The initial stage was largely limited to *relocation diffusion* (described earlier in this chapter). That is, large numbers of immigrants and officials moved from Britain to the colonies and, of course, took their language with them. Once there, they intermingled to different degrees with native peoples and non-English speaking immigrants, many of whom acquired English by *contagious diffusion* — contact with English speakers.

English now enjoys the status of *official language* — the one in which government business is transacted and printed, as well as the language of publicly financed education — in virtually all of Britain's former colonies. In many cases it is also the *vernacular language* — the one that is spoken by the people of a particular locality. But official and vernacular languages are not always the same in a given area or region. English, for example, is the official language of the United States and most Americans speak it, but literally millions of people living in ethnic neighborhoods, Indian reservations, and other enclaves across the land speak a different vernacular language. Look again at Figure 13-5 and you may get the impression that everybody in the United States, Australia, and New Zealand speaks English. Not so. People in parts of each country speak a different vernacular language.

Many countries that are former colonies have adopted the language of the colonizer as their official (or co-official) tongue even though, in many cases, only a minority of the populace speaks it. Examples include English in Ghana and Kenya; French in Senegal and Madagascar; and Portuguese in Angola and Mozambique. Typically, European languages are given official status in the former colonial realm for two reasons:

- ✔ The country contains numerous ethnic groups, some of which have a history of friction. Elevating one local language to official status could lead to jealousy and unrest on the part of other groups. Use of a European tongue favors no one and, in effect, puts everybody at an equal disadvantage.

- ✔ Use of a European tongue stands to promote international trade and commerce more than would a local language, which may be spoken nowhere else on Earth.

Nevertheless, many (even most, in some cases) of the native peoples in these countries continue to speak their own tongue as the vernacular language. In most cases, use of the official language(s) is concentrated in the cities and larger towns, while the vernacular persists in the smaller towns, villages, and rural areas. To the extent that everyday use of the official language is gradually "trickling down" from urban to rural areas, its spread exemplifies the process of **hierarchical diffusion**.

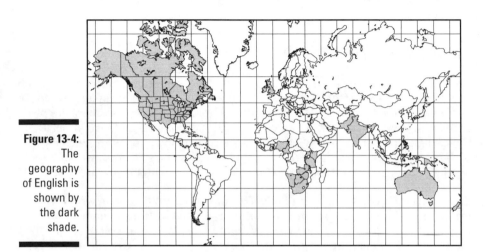

Figure 13-4:
The geography of English is shown by the dark shade.

Checking the physical effects

Language and physical geography may interact in various ways. The two most significant ways are through environmental terminology and linguistic refuges.

Environmental terminology

Languages tend to develop robust vocabularies that pertain to locally observed environmental conditions, and weak vocabularies that pertain to unfamiliar settings. English, for example, is weak in native terminology that pertains to deserts, the sub-arctic, very mountainous areas, and other characteristics that are not common to England. Thus, English has adopted environmental terminology from other languages to describe things that English cannot, or at least not very well. Accordingly, standard English dictionaries now include terms such as **arroyo** (from Spanish) to describe intermittent streams in desert environments, **taiga** (from Russian) to describe high-latitude coniferous forest, and **fiord** (from Norway) to describe steep-sided, glacially carved inlets of the sea.

Linguistic refuges

A **linguistic refuge** is an area where a language is insulated against outside change by virtue of remoteness, or the remains of a locale where a once widespread language continues to be spoken. Acting as physical barriers, aspects of the physical environment have served to isolate speakers of various languages and thus preserve their native tongues from outside agents of change. Heavily forested and extremely mountainous areas, as noted earlier in this chapter, have historically served that purpose.

The traditional Welsh and Irish languages, for example, at one time appeared to be on the brink of extinction, relegated to remote peninsulas, islands, and valleys in their homelands following the onslaught of English. However, nationalist aspirations and heritage awareness have led to campaigns to resuscitate these languages and promote their everyday use. Central to these efforts have been human resources — native language speakers — many of whom hail from villages and farms in linguistic refuge areas.

Playing the landscape naming game

Language may provide cultural character to the physical environment as well as to people. For example, what do New Jersey, Lake Okeechobee, Baton Rouge, and El Paso have in common? The answer is they are all *toponyms* or place names. People the world over have a habit of naming landscape features, be they mountains, hills, rivers, lakes, bays, seas, deserts, forests, cities, towns, streets . . . the list goes on and on. *Toponymy,* the study of place names, may provide diverse geographical insights. As per the four locales mentioned, toponyms may tell us something about where the settlers came from, who used to live here, and what language the settlers spoke. Toponyms may also tell us something about past religious distributions. Catholic settlers in North America, for example, had a propensity to bestow religious names on their settlements more so than Protestants, no doubt in part to solicit the protective favor of the Almighty in an often-difficult frontier setting. Thus, towns named for saints abound, especially in Quebec and California (San Diego, Santa Barbara, San Jose, San Francisco, and so on).

Place names may also provide philosophical insights. For example, about two centuries ago American culture was affected by the *Classic Revival,* which involved a new reverence of ancient Greece and Rome. One manifestation is the existence in Upstate New York of literally dozens of cities and towns that were named or renamed in accordance with the classical theme. Examples include Syracuse, Rome, Utica, Ithaca, and Romulus.

One of the most maddening things about toponyms is that they can be literally changed overnight, immediately rendering millions of maps and atlases out-of-date. The change of Burma to Myanmar and Zaire to Congo are fairly recent examples. Prior to its dissolution, the USSR contained an estimated 20,000 places named for Stalin — mountains, cities, alleys, you name it (literally). When Stalin's legacy suddenly fell out of favor, so did toponyms in his honor. Few remain.

Creating a Single Global Culture

Is the number of cultures on Earth today increasing or decreasing? Will the cultural mosaic of your grandchildren's world have more pieces or fewer pieces? That is, will the world map contain more culture areas or fewer culture areas than exist today? The search for answers reveals powerful opposing sets of social forces.

Promoting cultural divergence

Cultural divergence is the concept and process of culture creation. I have shown how this works by virtue of isolation, adaptation, and barrier effects, as well as by the just-mentioned processes of nationalistic aspirations and heritage preservation. Because of modern means of communication, the first three of these are not nearly as powerful as they once were, though they are still players in culture creation. At the same time, cultural fads and technology changes are occurring with much greater frequency than ever before, and help to create peoples who are culturally different from neighboring groups. Consider also the human spirit that rejects a world in which everyone looks and acts the same. Cultural divergence may have slowed, therefore, but no evidence suggests that a world of 15,000 culture areas will reduce to a single culture area any time soon.

Promoting cultural convergence

It seems pretty clear, however, that in the present day and age more cultures are dying out than getting born. As many as a thousand languages are likely to disappear in your lifetime because they are not being spoken by the next generation who are instead learning and speaking another language. This is an example of *cultural convergence,* the concept and process of culture destruction and attendant "convergence," or coming together of the world's people in the cultural sense.

This is largely happening because communication is overwhelming the traditional factors that isolated people and encouraged cultural differentiation. The key players in this trend are improved transportation (especially the advent of road-building in remote areas); expansion of television, movies, and the Internet; increased literacy and access to printed media; rising tourism due to increased disposable income and leisure time; rising foreign trade; the growing number of multi-national corporations; and desire for economic development.

In the future, culture will continue to be the spice of life and place, even if the range of flavors is significantly less than we have today.

Chapter 14

Where Do You Draw the Line?

*Y*ou can divide up space and set up different types of boundaries in many ways. In addition to living in the United States, I live in the State of New York, the City of New York, and the Borough of Manhattan (also known as New York County), all of which have separate boundaries. I also live within a particular school district, police precinct, City Council District, U.S. Congressional District, New York State Senate District, and New York State Assembly District. They, too, have boundaries, and none are the same. But that's just me. What about you? How many political lines on the map enclose your residence? And the government isn't the only one that has this "thing" about dividing space. The phone company has divided the country in area codes. The postal service has done the same, only their lines have different placements and define different areas — zip codes. Maybe you belong to a church that divides space into dioceses, presbyteries, or something else. Maybe you work for a company that divides space into sales territories. The bottom line is that Earth is divided by lines of all kinds.

As all of these instances testify, *Homo sapiens* have proven to be a rather territorial species (as illustrated by the wall in Figure 14-1). Indeed, one of the most profound things we humans do to our earthly home is divide it up among ourselves, thereby creating regions of sovereignty, control, or admin-istration. We do this by drawing lines on a map, which begs a rather basic question: Where do you draw the line? This is a central issue of political geography, which is concerned with division and control of Earth's surface and is the subject of this chapter. In writing this chapter, I, too, had to draw the line — that is, make decisions regarding what to include and what not to. In the end, I decided to focus on two topics, international borders and voting district boundaries, which I believe are particularly timely.

Figure 14-1:
The Great Wall of China is an enduring symbol of the human fondness for dividing and controlling the Earth.

Drawing and Re-Drawing the Boundaries of the World

Many years ago I attended a lecture concerning "lost nations." I had no idea what it was going to be about. I figured maybe the entire populations of several countries had gotten off at the wrong exit, refused to ask for directions, and were, you know, lost. Or maybe they were all stuck on a subway someplace, like that old folk song about the man who never returned. I was wrong. The presenter had prepared a geographical analysis of all the countries of the world that had ceased to exist since 1850. Dozens and dozens of those countries had been "lost" to the family of nations. Maps of the world's countries are quite common. But neither the nations nor the boundaries between them are etched in stone. Since that lecture, the Soviet Union, East Germany, Czechoslovakia, and maybe one or two other countries have joined the "lost" list. But the world is hardly running out of nations. Indeed, new countries seem to be springing up faster than existing ones are disappearing. The Soviet Union fractured into 15 nations and Yugoslavia into 5. Czechoslovakia split into the Czech Republic and Slovakia. In Africa, Eritrea broke away from Ethiopia.

Meanwhile, a host of new candidates are vying for countryhood. People in Chechnya want to be free of Russia. The Tamils want to be separate from Sri Lanka. Basques in Spain and France want a country of their own. And on it goes. If past is prologue, then I have no doubt that a comparison of today's world map with future ones will reveal a mix of new nations and lost ones.

Tensions over turf, like debt and taxes, never seem to go away. For starters, some ethnic group is always clamoring for independence or wanting to draw or re-draw boundaries in a certain part of the world. Without taking sides in such matters, let me simply propose a singular reason for much of this contentiousness. Namely, the people who drew most of today's international boundaries came from somewhere else, and they often gave little or no thought to the possible long-term consequences to the native population.

Typecasting Boundary Lines

Insight into the nature and consequences of boundary making is provided by Figure 14-2. Assume it is a map from more than a hundred years ago that shows the boundaries of five ethnic groups labeled A–F, as well as a river and a line of longitude. Also assume that you are an official who, in the language of the previous paragraph, comes from somewhere else, and must decide where to place an international boundary within the area shown in Figure 14-2. Three options are available, and they correspond to the features just mentioned. They are:

- ✔ **The Ethnic Option:** The homelands of five ethnic groups, indicated by the letters A–F in Figure 14-2, occupy the area in question. One option is to divide the area on the basis of the ethnic boundaries highlighted on the map.

- ✔ **The Landform Option:** A fairly large river flows through the area. A line running down its middle could be used as the border.

- ✔ **The Graticule Option:** If you read Chapter 3, then you may recall that *graticule* refers to Earth's grid of latitude and longitude. A line of longitude runs through the region. This line potentially could serve as the international boundary.

I didn't simply dream up these possibilities. Instead, the three options represent the three kinds of borders that political geographers most commonly identify. Each type is discussed in greater detail in the following sections.

Ethnic boundaries

This one's a no-brainer. Ethnic boundaries separate the homelands of neighboring ethnic groups. They are also probably the oldest kind of boundary there is. Viewed against the broad sweep of human history, the concept of a country with recognized borders and a centralized government is a fairly recent invention. Much older is the idea of an ethnic group possessing a distinct language, customs, and homeland.

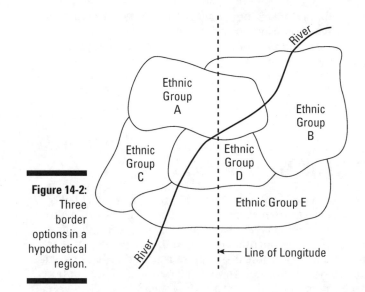

Figure 14-2:
Three
border
options in a
hypothetical
region.

Ethnic Group A

Ethnic Group B

Ethnic Group C

Ethnic Group D

Ethnic Group E

River

River

Line of Longitude

For example, when the Europeans arrived in North America, they didn't find a country called the United States divided by lines into entities named New Jersey, Kansas, Arizona, and so forth. Those came much later. What they did encounter, however, was a continent inhabited by dozens upon dozens of Native American ethnic groups, each of which had its own distinct homeland rather like the areas labeled A through F in Figure 14-2. Similarly, when Africa was about to be divided into colonies, it did not consist of the countries we see today, but instead was composed of the homelands of hundreds of ethnic groups.

Typically, the boundaries of ethnic homelands in North America, Africa, and elsewhere were not fixed lines established by formal treaty. More often they were ill-defined zones that lay on the periphery of lands occupied by an ethnic group. Thus, they were fluid — prone to change. That is, if a particular ethnic group gained or lost territory as a result of some conflict, the ethnic boundary changed accordingly. For these two reasons — ill definition and impermanence — ethnic boundaries were rarely used as a basis for modern boundary making. Instead, the treaty makers who were responsible for most of today's boundaries opted for the types discussed in the next two sections.

Natural (physical) boundaries

Natural boundaries are based on some aspect of *physiography* — physical landforms. While diverse aspects of landscape may be called upon, three are particularly common. The United States provides abundant examples of each type (see Figure 14-3):

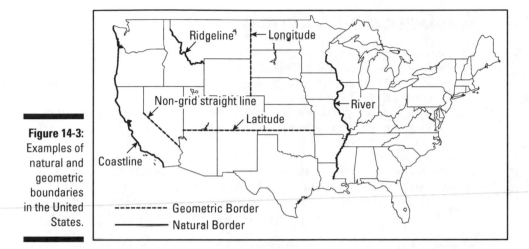

Figure 14-3:
Examples of natural and geometric boundaries in the United States.

✓ **Shorelines:** The Atlantic Coast boundary of Florida and the maritime borders of Alaska are good examples. The shorelines of the Great Lakes States are relevant to some extent, though the border with Canada runs through the lakes rather than following the shoreline.

✓ **Rivers:** The New Jersey-Pennsylvania border (Delaware River); the Illinois-Iowa border (Mississippi River); and the Oregon-Washington Border (Columbia River). Quite typically, the border is not the left bank or the right bank, but instead an imaginary line that runs down the middle of the waterway.

✓ **Ridgelines:** The crest of a ridgeline in the Bitterroot Mountains marks the border between Idaho and Montana. Similar crests in the Appalachian Mountains mark the boundaries between Virginia and West Virginia, and between North Carolina and Tennessee.

The principal advantage of natural borders is that they are based on things that can be readily seen, agreed upon, and mapped. In contrast to an occasionally questionable ethnic boundary, there should be no question as regards to the location of a river, coast, or ridge. They are not necessarily permanent, however. As you can see in Chapters 6 and 7, powerful tectonic and gradational forces build up and wear down Earth's surface over time. Rivers in particular may change course as a result of a major flood or cataclysmic event. But the advantages of natural boundaries are powerful. As a result, many boundaries are based on this feature.

Geometric boundaries

Geometric boundaries are straight lines. In a majority of instances they are lines of latitude and longitude. Several examples are evident on a map of the United States, including the latitude line that separates Utah, Colorado, and Kansas from Arizona, New Mexico, and Oklahoma (see Figure 14-3). Some straight lines are not part of Earth's grid, however. Perhaps the most noticeable U.S. example is the northwest-southeast line that constitutes most of the border between California and Nevada (see Figure 14-3).

Use of straight lines is not limited to the United States. Nearly all of the borders between the states of Australia are lines of latitude and longitude. In West Africa, similarly, the country of Mali has both a latitudinal and longitudinal border with Mauritania, and a non-grid straight-line boundary with Algeria.

The most attractive feature of geometric boundaries, as far as border-markers are concerned, is that they are based on abstract geometric space instead of the physical Earth or ethnic homelands. Therefore, they have a degree of permanency that the other types do not. Let it rain for 40 days and 40 nights, and the results may alter natural boundaries, but geometric boundaries will not be affected. Similarly, let every ethnic group on Earth riot, have a bad hair day, go on the warpath, or whatever, and the results may alter previous ethnic boundaries; but again, geometric boundaries will be unaffected because they are abstract concepts. People can't pick up and move a line of latitude or longitude. They could conceivably "fake the location" of such a line for personal benefit, but proper surveying can detect and fix such shenanigans.

Living with the Consequences

After the boundary lines have been drawn, the citizens of those areas and the ones responsible for creating the lines have to live with the consequences. The world's international borders consist of thousands of linear segments that enclose some 200 countries. In some cases the results have proven functional and harmonious. Other cases, however, are characterized by very different adjectives. Following are an overview and examples of some of the more challenging consequences of global boundary making.

Ethnic intrigues

As I explain earlier in this chapter, much of the world is divided on the basis of natural and geometric boundaries as opposed to ethnic ones. This is especially true of the many countries that once were colonial possessions. Over large parts of Earth, therefore, disparities exist between ethnic boundaries and country boundaries. Visual comparison of these border types in Africa, for instance, reveals two very separate sets of lines that seldom coincide. Though not necessarily bad, disagreement between these sets of lines has occasionally fostered significant ethnic and political tension in Africa and elsewhere.

This unrest broadly falls into three categories. Before discussing them, however, I need to make a particular distinction between state and nation. For the purposes of this discussion, *state* is synonymous with country. It refers, therefore, to territory that is delineated by a border and has sovereign, independent status. *Nation,* in contrast, is synonymous with an ethnic group that has its own traditional homeland, history, and culture that causes its members to view themselves as distinct from other people. Sometimes, for example, particular Native American peoples are referred to as the Seminole *Nation,* the Hopi *Nation,* the Sioux *Nation,* and so forth. That is the meaning of "nation" used here.

Multi-nation states

A *multi-nation state* is a country that contains the homelands of more than one nation. Nigeria, whose population consists of the homelands of some 200 national (ethnic) groups, is a prime example. The main problem in this situation is trying to create a common identity and sense of unity among peoples who speak different languages, practice different religions, and, perhaps worse, have a history of mutual animosity, if not outright hostility. Consider the following scenarios, any one of which might result in a level of ethnic tension that results in unrest:

- ✔ Appointed government officials, top military leaders, or chiefs of police are members of an ethnic group that is distrusted by members of other ethnic groups.

- ✔ Political parties form along ethnic lines, heightening prospects that election results may lead to friction between nations.

- ✔ Government declares official languages or religious holidays that promote the culture of one or more nations but not others.

- ✔ The capital is located in (or relocated to) the homeland of a particular nation that is distrusted by others.

- ✔ Economic development projects are disproportionately located in the homeland of a particular nation, leading other groups to feel they are being deprived of their fair share of economic opportunity.

Multi-state nations

A *multi-state nation* consists of an ethnic group whose people and traditional homeland lies within two or more countries. This situation may lead to *irredentism,* a foreign policy in which one country seeks to acquire foreign territory for the purpose of unifying a nation. The term dates from the 1870s when Garibaldi and his followers sought to incorporate "unredeemed" *(irredenta)* Italian-speaking territories into the Italy that you see on today's map. The potential problem, of course, is that a country that contains the coveted land is not likely to give it up without a fight.

The Somalis provide a case study. This ethnic group's traditional homeland includes much of the "horn" of East Africa (see Figure 14-4). In the latter part of the 19th century, Ethiopia, Italy, and Britain signed a treaty that divided this region between them. Following de-colonization, the present Somali Republic was established out of the territory controlled by Italy and Britain. But a large portion of the traditional Somali homeland, known as the Ogaden, remains under Ethiopian control.

Figure 14-4:
The horn region of East Africa.

Somali politicians have argued that dividing up their traditional homeland among foreigners without Somali consent was not fair in the first place; thus the Somali Republic rightfully ought to annex Ogaden by any means necessary. Naturally, this argument echoes rather strongly among Somalis but alarms Ethiopia, which views the Ogaden to be part of its territory by virtue of a legitimate treaty. The result has been on-again, off-again debate and occasional armed conflict between the two countries over control of the Ogaden.

State-less nations

A *state-less nation* is an ethnic group whose traditional homeland is currently under the control of another country. Virtually hundreds of ethnic groups and their homelands are in this category. These peoples usually adopt one of three stances:

The Kurds

The Kurds, who number some 20 million, are an example of both a multi-state nation and a state-less-nation. In the early years of the 20th century, their traditional homeland, Kurdistan, lay mainly within the Ottoman Empire, which ceased to exist after World War I. Kurdish nationalists hoped that an independent Kurdistan would be born of post-war boundaries drawn by the victorious parties. Sadly for them, that did not come to pass. Instead, the modern map of the region gradually emerged, in which Kurdistan is now spread over six countries (hence, a multi-state nation), principally Turkey, Syria, Iraq, and Iran. The Kurds are not a majority in any of these countries, but they have not given up hope of independence for their homeland. Until that happens, however, they remain a state-less nation.

- ✔ **Do nothing:** A majority of stateless nations are comparatively docile about this particular circumstance. While some members may desire statehood, the overall level of support for autonomy has not — for whatever reason — reached a critical level.

- ✔ **Seek greater autonomy:** Some ethnic groups seek a greater degree of autonomy for their homelands as opposed to complete independence. The Welsh and Scottish peoples are examples. Although some hard-core nationalists long for complete separation from the United Kingdom, the majority favor continued union complemented by a high degree of self-rule.

- ✔ **Clamor for independence:** Members of some ethnic groups seek complete independence of their homelands from the country to which they are presently bound. People in predominantly French-speaking Quebec, for example, have sought sovereignty for their province and taken the issue to the ballot box, where it failed to pass. In other cases, however, people are so aggrieved by the perceived injustice of their circumstance that they feel any means of achieving their goal — including extremely violent acts — are justified. Examples of such groups include the Basques, Tamils, and Kurds.

Positional disputes

Positional disputes concern disagreements over location of a border. One of three factors typically underlies the resulting tensions.

Poor definition

In some cases, disagreement arises between neighboring countries concerning borders (often inherited from colonial times) that were poorly defined. Here are two examples:

- **Kashmir:** At the end of the colonial era, a boundary commission sought to divide this Himalayan region between Pakistan and India, taking into account ethnic divisions. Nationalists on both sides disagreed over the recommended partition of the region, which is also the source of rivers that supply irrigation water to the countries' agricultural lowlands. Control of Kashmir continues to be a major bone of contention in Indian-Pakistani relations.

- **The Southern Andes:** During Spanish colonial administration, ridges and watersheds were used to define what would become the long border between Argentina and Chile. The area was poorly mapped, however, resulting in conflicting interpretations of boundary locations. Segment by segment, disputes over different parts of the boundary were resolved by direct negotiations, third-party mediation, the International Court of Justice, or — in one particular case — the intervention of the Pope

Invisibility

In some cases, determining where a border is located is impossible. Such situations are potential powder kegs. Here are two examples:

- **Amazonia:** The periphery of the Amazon basin contains a couple of boundaries that pass through trackless expanses of jungle and rainforest. Thus, while the line is plainly visible on the map, no clue to its whereabouts can be found on the ground. The potential for trouble has been highlighted by discovery of oil and natural gas in the region.

- **Arabian Peninsula:** Because of the vast emptiness of the Arabian Desert, the boundaries between Saudi Arabia and Yemen, Oman, and the United Arab Emirates have been approximated but not delineated. Thus, on most detailed maps of Arabia, the borders are shown as dashed lines accompanied by "boundary undefined" or words to that effect. So far the countries have felt no major ill effects, but situations like these have a way of turning nasty.

"Acts of God"

An "act of God" is a cataclysmic event that results in disagreement over the location of a border. The most common (but hardly everyday) occurrence involves a boundary river that changes course following a major flood. The Mississippi River, for example, has changed course several times since it was established as the boundary between Arkansas and Mississippi. As a result, some land that used to be on the Arkansas side of the river is now on the Mississippi side, and vice versa. Similarly, a small portion of Texas ended up on the Mexican side of the Rio Grande years ago following a freak flood.

If a parcel of land "changes sides," then is it in the same jurisdiction it was before, or in a different one? Basically, it all depends on how the treaty was written. In any event, however, you can no doubt appreciate the potential here for political intrigue.

Functional disputes

Functional disputes concern policy disagreements regarding immigration, trade, customs fees, or some other matter(s) that apply to a particular boundary. Here are two contemporary examples.

United States-Mexico

Trade and daily exchanges across the U.S.-Mexican border are at an all-time high and have served to bring the two countries closer together than ever before. At the same time, however, illegal immigration and drug trafficking have caused many Americans to advocate a "hardening" of this border to make crossing more difficult. Thus, contrasting forces are respectively advocating a more open boundary and a less open boundary. On the American side, border patrols have been increased and physical barriers erected along some stretches. Critics view these physical barriers as virtual Berlin Walls, while proponents view them as necessary steps to maintain national security. Mexicans tend to view America's "hardening" policies as counterproductive to good relations and often contrast it to the U.S.-Canadian border, which is very open and largely non-policed.

Israel-Palestine

Significant exchanges occur across borders that separate Israel from land administered by the Palestinian Authority. Overwhelmingly, this involves Palestinians who commute to work in Israel or who must travel through Israel to reach other lands under Palestinian control. From the Palestinian perspective, an open border is desirable. However, Israeli security concerns have caused Israel to take a different view of these same boundaries and, during times of strife, close them entirely as a means of expressing displeasure or retribution.

Resource disputes

Occasionally an international boundary cuts across a natural resource whose use becomes a bone of contention between the neighboring countries. Here are two examples.

The Rumaila Oil Field

This oil field straddles the border between Kuwait and Iraq, as shown in Figure 14-5a. Though it lies principally within Iraq, the nature of oil means that wells drilled in Kuwait can potentially suck up this resource as fast or faster than wells across the border. Sensitivities regarding "fair share" have run high. Indeed, increased production by Kuwait was one of the rationales offered by Iraq for invading its neighbor in 1990.

The Georges Bank

For many years, this historically rich fishery, shown in Figure 14-5b, straddled a disputed marine boundary between the United States and Canada. Like the preceding example, each country had access to it and was sensitive to the notion of "fair share." The fish, who could care less about the boundary's location, habitually swam back and forth across the disputed area. That meant that over-fishing on one side of the border, while relatively enriching that country's fishermen, had the potential to deplete the entire fishery and, at the same time, deprive the neighbor of "fair share." Fish bones of contention? You bet, and hold the tartar sauce.

Figure 14-5 a and b: The Rumaila Oil Field and Georges Bank

Sometimes resource disputes are resolved amicably by direct discussion between the parties involved, or with the assistance of an organization with which both are affiliated — like the UN or NATO. Sometimes the case literally goes to court — the International Court of Justice, that is. The dispute between the U.S. and Canada over the Georges Bank was settled in that manner. On the other hand, sometimes the problem simply festers or is resolved by armed conflict. Thus, resource disputes have diverse outcomes — some peaceful, some not.

Land-locked states

A *land-locked state* is a country that is completely surrounded by land that belongs to one or more neighboring countries. Thus, it has no port of its own and therefore no immediate access to international waterways. Three dozen land-locked states are located throughout the world, including about a dozen each in Africa and Asia, two in South America, and the rest in Europe (see Figure 14-6). At least three potential problems emerge from this status.

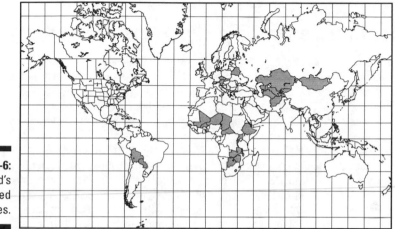

Figure 14-6:
The world's land-locked states.

Inhibited access to foreign markets

Lack of an ocean boundary carries with it lack of unfettered access to foreign markets. You may be thinking, "Who needs a sea port if you have an airport? Why not import and export everything by airplane?" Aside from the fact that you can't just go flying over another country's territory without permission, transporting bulk trade items by ship is far and away the cheapest mode of importing and exporting goods. For all intents and purposes, therefore, access to foreign markets requires access to the high seas.

In recognition of the peculiar position of land-locked states, international law has long granted them the *right of innocent passage* in matters of foreign trade. That is, land-locked countries have the right to transport imports and exports across the territory of neighboring countries provided such action does not pose a threat to the security of the host country or violate its laws. A land-locked state may even be allowed to lease dock facilities in a neighboring country, thereby giving it the port it would otherwise lack.

Treading softly in foreign relations

Land-locked states find it beneficial to tread very softly in foreign affairs, especially in matters that pertain to neighboring countries that grant them the right of innocent passage. Tick off thy neighbors, and they may cut you off from the sea, bringing your ability to import and export to a halt. Obviously, the effect of this type of action on your economy may be devastating. So make nice, even if you despise them. And you can scream all you want to about your rights. It turns out that the vaunted right of innocent passage is little more than some touchy-feely gentleman's agreement. Or as they may say in New York, the right of innocent passage plus $1.50 gets you a ride on the subway.

Applied Geography: Flying over Pakistan

Following the events of September 11, 2001, the United States military lent its considerable support to the overthrow of the Taliban-led government of Afghanistan. As part of that effort, numerous air strikes and cruise missiles were launched from American warships operating in the Arabian Sea. Also, a sizeable contingent of marines and other personnel were flown into Afghanistan from the same flotilla. But because Afghanistan is land-locked, all of those sea-launched flights had to fly over some other country to get to their destinations, and that country was Pakistan.

Countries have sovereignty over the "air space" (that is, the physical atmosphere) that is directly above their territory. Flying through that air space without permission is tantamount to invading or violating the sovereignty of the country in question. It is a very serious matter. Accordingly, in the days and weeks prior to the beginning of hostilities in Afghanistan, U.S. government and military officials carefully courted their Pakistani counterparts to join the coalition against Afghanistan. This assistance was to prove important for several reasons, one of which was to permit important military missions that involved Pakistani airspace.

Increasing the cost of trade

Even if all is well with respect to the previous two points, being land-locked means that your imports and exports absorb extra transportation costs versus countries with a coast. That increases the cost of its exports (and therefore makes them less competitive in global markets) and raises the cost of imports, too. Neither is beneficial to an economy.

Questions of size and shape

I can hear the "tsk-tsking" already, but I'm going to say it anyway. Does size matter? How about shape? That is, are some sizes and shapes better than others as far as the welfare of a country is concerned? Clearly, countries come in all sizes and shapes.

Russia, the biggest country on Earth, contains nearly 6.6 million square miles. On the other end of the spectrum lies Monaco, which barely covers 1 square mile. Lots of other sizes can be found in between. Logically, we all probably assume that bigger is better. More territory usually means greater prospects for mineral wealth, more agricultural land, and so forth. But it may also mean remote regions, lots of boundary to defend, restless ethnic groups, large uninhabited areas, and umpteen challenges to cohesiveness. A small, compact country may face none of these challenges.

And that leads to a consideration of shape. Political geographers tend to agree that shape clearly can make some kind of difference in some cases. The following sections discuss the four types of country shapes, which are shown in Figure 14-7.

Figure 14-7: Different shapes of countries.

Compact states

A *compact state,* such as Uruguay, is roughly circular and is generally thought to be the ideal shape. Assuming absence of mountain ranges or other physical barriers, parts of the country may be readily interconnected with compara-tively little investment in roads, railways, electronic cables, and so forth. Also, compactness minimizes the amount of international border that may need to be defended, and may also minimize the extent of internal ethnic boundaries as well. All in all, compactness tends to promote internal cohesion.

Fragmented states

A *fragmented state* consists of scattered, disconnected pieces of territory, typically because all or part of the country is a chain of islands. Indonesia is an example. Fragmentation makes it more difficult for government to impose central control and to promote cohesion and interaction between different parts of the country. Also, different fragments may be homelands of different ethnic groups that have little in common.

Elongated states

An *elongated state,* such as Chile, is long and thin. In most cases this shape is detrimental to national cohesion. Some places may be remote from the capital, a lot of border needs to be defended, and elongation often results in a more diverse populace than a compact shape. On the other hand, elongation may result in climatic variety that is a plus to the country's agriculture and economy.

Prorupted states

A *prorupted state,* such as Namibia, is generally compact but has a noticeable protrusion. In the case of Namibia, a land protrusion gave Germany (once the colonial ruler in Namibia) access to the Zambezi River. In the cases of other countries, such as Thailand and Myanmar (Burma), proruption is the result of a peninsula. This shape is generally thought to be disadvantageous because it isolates a portion of the country.

Drawing Electoral District Boundaries

A very different, yet very important, component of political geography concerns the drawing of voter district boundaries. The governments of the United States, Great Britain, Canada, Australia and other democracies include legislative bodies elected on the principal of "one person, one vote" — equal representation for all citizens. This in turn necessitates creation of electoral districts that contain roughly the same number of people. Regarding the United States, drawing the boundaries of Congressional Districts immediately comes to mind, and so, too, the outlines of other political geographic entities — such as city council districts, state senate districts, and state house districts. Making "one person, one vote" a geo-political reality involves four steps:

1. **Count:** Conduct a nationwide census that is as accurate as possible. The United States does one every ten years. Indeed, the Constitution requires it — not simply to count Americans, but for the specific purpose of acquiring data that permits creation of equal Congressional Districts.

2. **Map:** The next step is to map the collected information with as much geographic accuracy as the data permits. That means, for example, that when it comes to a large city, boundary-makers need population statistics on a block-by-block basis.

3. **Calculate:** The next step is to calculate the number of people who should be in each district. This is a fairly simple matter. Say, for example, that you are considering a given state's house of representatives that has 100 seats. In that case, you take the state's population total, as revealed by the census, and divide by 100. The numerical result is the number of people who *should* reside in each district.

4. **Draw:** The final step is to draw the map. In the case of the above, that means dividing the state in question into 100 polygons, each of which (as closely as possible) contains the same number of people, as calculated in the previous step.

Gerrymandering: Rigging the outcome

"Figures don't lie, but liars do figure." Maybe you've heard this cute little phrase, which alludes to the possibility of using or manipulating data to serve a particular interest. As regards political geography, it may manifest itself in *gerrymandering,* the drawing of voter district boundaries to benefit a particular group or political party.

Gerrymander pays mocking tribute to Elbridge Gerry, who, as governor of Massachusetts in 1812, sought to redraw the political district map of his state in such a way that assured the continued dominance of his political party in the statehouse. One particular political district was drawn in a narrow, sinuous shape that, according to one observer, "looked like a salamander." That prompted a wag to respond that it was no salamander, but instead a "gerrymander." The moniker stuck.

Then and now, gerrymandering typically occurs because a political party or voting bloc wants to maximize its representation in a legislative body and minimize that of another group. Figure 14-8 shows the two most common types of gerrymanders. By way of background, each diagram is characterized by the following:

- ✔ The area in question consists of a city (inner circle) and its surrounding countryside.

- ✔ An accurate census has placed the region's total population at 50,000.

- ✔ Two categories of voters live in the district, the As and the Bs. You may think of them as members of different political parties or different racial or ethnic groups or whatever. The key thing is that they are on opposite sides of the political fence.

- ✔ Every member of Group A lives in the city. Every member of Group B lives outside it. (This is a gross simplification of typical voting bloc geography, and is made strictly to facilitate understanding of gerrymandering techniques.)

✔ Representatives of Group B are currently in power. They will resort to any legal means to stay there, including drawing voting district boundaries designed to promote election of their candidates.

✔ The area in question is to be divided into 5 political districts, each of which contains 10,000 people.

"Diluting" a voting bloc

In Figure 14-8a, Group A consists of 20,000 people and Group B 30,000. Because Group A accounts for two-fifths (40 percent) of the voters, you may expect them to capture two-fifths of the seats being contested. That is, they may win two of the five seats. But they can quite possibly get nothing. In Figure 14-8a, members of Group A have been spread out over the five districts such that they are in the minority in every instance. In a manner of speaking, their voting power has been "diluted." Assuming voters vote solely along party lines, the results will be 5-0 in favor of Group B.

"Packing" a voting bloc

In Figure 14-8b, Group A consists of 30,000 people and Group B 20,000. Because Group A now accounts for three-fifths (60 percent) of the voters, you may expect a totally different electoral outcome. But there is a way to draw the district boundaries that ensures political dominance for Group B. And because these are the people who are drawing the lines, you can bet they will do it. In the diagram, two inner-city districts consist entirely of members of Group A. But by "packing" so many members of Group A into these two districts, the result is that they are in the minority in the other three. Assuming once more that voters vote solely along party lines, the results will be 3-2 in favor of Group B. Stated differently, the bloc that consists of two-fifths of the voters will control three-fifths of the seats.

Figure 14-8 a and b: These diagrams show two kinds of gerrymandering: diluting and packing.

Meeting the letter and spirit of the law

You may be thinking, "That's not fair!" And you may be right. But more important is the question, "Is that legal?"

Some people say "Yes." Clearly, every political district in Figure 14-9 complies with the letter of the law. That is, the number of voters is the same in each district, so the constitutional concept of equal representation has been met.

Other people, however, believe that elected officialdom should be as diverse as the represented population. In their view, gerrymanders violate the spirit of the law, lead to a distrust of government, and may ultimately undermine the very fabric of American democracy.

A practical solution, in their view, is to create voting districts drawn to virtually guarantee the election of members of particular ethnic groups (Figure 14-9). Opponents view this solution as blatantly unconstitutional and say districts drawn to encourage election of particular peoples are as offensive as lines drawn to exclude them.

Figure 14-9:
An oddly shaped congressional district drawn to encourage election of a minority group member to Congress.

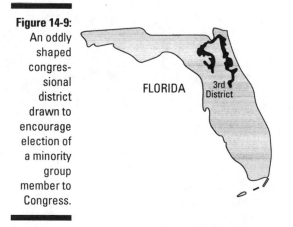

The courts, for their parts, have flip-flopped on the issue. Quite literally, the jury is still out. In the meantime, passions run high among folks who feel strongly about this issue, which is as much about geography as it is politics and law. After all, the question comes down to "Where do you draw the line?"

Part IV
Putting the Planet to Use

The 5th Wave   By Rich Tennant

It took last week's big rain for us to really appreciate having a deck.

In this part . . .

The Earth is more than our home. It's also the source of our daily bread. Not to mention our daily aluminum can, newspaper, job, gallon of gas, and everything else that sustains us. And at the end of each day, we take out the trash, which gets picked up by a big truck and goes . . . where?

Humans cannot help but interact with Earth's surface. On it we plant our crops, locate our places of business, and build our settlements and cities. From it we draw a host of resources that we either consume as they are, or transform into a host of products that define the various global economies and standards of living. And at the end of the day, we take out the trash. In doing all of these things, we sometimes modify our earthly environment in ways that pose significant challenges for present and future generations.

In this part, you will learn some of the key concepts and phenomena that concern human use and misuse of planet Earth. Economic geography, resource geography, urban geography, and environmental geography will be the specific topics. All have a central bearing on this Part's central theme — how humans interact with and impact the only planetary home we are likely to know.

Chapter 15

Getting Down to Business

· ·

In This Chapter

▶ Finding different ways (and places) to make a living

▶ Choosing the best business location

▶ Navigating transportation systems and networks

▶ Moving to cyberspace or real space?

· ·

When I was about 10 years old, somebody bought me a geography board game in which you scored points by correctly identifying the key products of different countries. I don't remember how it was played, but certain facts were seared into my memory. Honduras = bananas. Australia and New Zealand were heavily into sheep, and India produced jute big time. Mind you, I had no idea what jute was, but I knew that it came from India and that was what mattered!

Over the years I have met several people who not only have a head full of similar facts (Argentina = beef, South Africa = diamonds), but also equate knowing them with knowing geography. These folks are not totally misguided. *Economic geography,* the subject of this chapter, is a major subfield of geography and concerns the location of different means of livelihood. And indeed, for a long time, economic geography was largely concerned with knowing where products came from, and why.

In recent decades, however, economic geography literally has been getting down to business. That is, though still concerned with the distribution of economic activity, economic geography has acquired a focus on optimal business locations and strategies that enhance business success. In so doing, it has come to embody quite strongly the two modes of geographical thinking: analysis of where things *are* located, and where they *should be* located.

This chapter addresses both perspectives with emphasis on where things *should be* located.

Categorizing Economic Activity

Economic activities are all of the ways that people the world over make a living. Obviously, a complete listing plus descriptions would be encyclopedic. To simplify matters, and as a first step toward geographical analysis of livelihoods, geographers commonly characterize economic activities as a four-part scale whose components — primary, secondary, tertiary, and quaternary activities — become progressively more complex and detached from the natural environment.

Primary activities

Primary activities involve harvesting the Earth or extracting raw materials from it. Examples include agriculture, forestry, fishing, and mining (Figure 15-1). In each case, the relationship with the environment is intimate and location-specific — meaning that you can't grow bananas, catch tuna, or mine coal just anywhere. Instead, you can only extract resources where they occur or, in the case of agriculture, grow particular crops where conditions are right. As a result, location options for many primary activities are limited.

Secondary activities

Secondary activities change the form of raw materials in ways that add to their value. Examples include processing and manufacturing. Thus, a cheese maker may transform milk into cheddar, which has a different form and higher price than its raw material. Similarly, an automobile manufacturer might bring together headlights, tires, upholstery, spark plugs, and so forth at an assembly plant and transform them into a vehicle that has a different form and higher cost than the sum of its parts.

As these examples illustrate, the raw material required by a secondary activity could be either a natural substance (milk) or products of previous processing or manufacturing (tires). Those secondary activities that require natural raw materials usually are located close to relevant primary activities in order to minimize transportation costs. In contrast, those secondary activities that require processed or manufactured raw materials are less intimately tied to the natural environment and have more location options.

Tertiary activities

Tertiary activities provide goods and services. They include retailers and wholesalers, who link producers and consumers; and personal and professional workers, including clerical personnel, financial services, entertainers, lawyers, and doctors. Tertiary activities typically have no direct relationship to environmental items (such as farming and fishing) and therefore have few constraints on possible locations.

Figure 15-1:
This open pit gold mine in Australia is an example of primary economic activity.

Quaternary activities

Quaternary activities involve research, information processing and dissemination, and organizational administration. "White-collar professional" probably summarizes this category as well as any single term can. Figure 15-2 shows an example of white-collar professionals doing a quaternary economic activity. Typically, these activities are urban-based and completely detached from the natural environment.

Figure 15-2:
These IRS data processors personify quaternary economic activity.

Activity distribution around the world

The relative presence of these four categories of economic activity varies geographically around the world. Generally, the economies of the least developed countries tend to have a very high percentage of their workforces engaged in primary activities, and progressively lesser percentages for the others (as shown in Figure 15-3). As national economies become more developed, the workforce tips toward the higher end of the range. And as a result, the most advanced economies tend to have a high percentage of their workforces in the tertiary and quaternary categories, and much smaller figures for the primary and secondary groups.

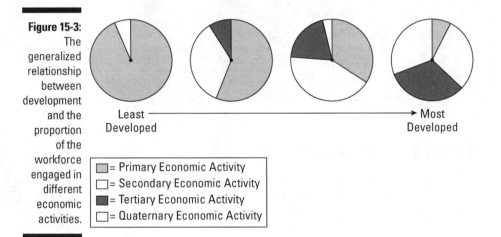

Figure 15-3: The generalized relationship between development and the proportion of the workforce engaged in different economic activities.

Least Developed ⟶ Most Developed

☐ = Primary Economic Activity
☐ = Secondary Economic Activity
■ = Tertiary Economic Activity
☐ = Quaternary Economic Activity

Also, while these four categories of economic activity involve separate and distinct means of livelihood, they are nevertheless linked, especially in the more advanced societies. Thus, tuna caught by the crew of a trawler (primary activity) may be offloaded at a factory for processing and canning (secondary activity). Wholesalers then purchase large quantities of canned fish and distribute them to local grocery stores for sale (tertiary activity). Overseeing this buying and selling, as well as attendant marketing, are company executives (quaternary activity).

Putting Economic Systems into Place

In contrast to economic activities, which may be viewed as individual livelihoods, *economic system* refers to the organization and functional nature of a national economy as a whole. *Subsistence economies,* which are prevalent in the less developed countries, and *commercial economies,* which are prevalent in the more developed countries, are the two principal types of economic systems. Technically, a third category, *planned economies*, is characterized by government control of supply, demand, prices, employment, and distribution. With the collapse of most Communist societies, however, this has ceased to be a significant player on the world stage, at least for the time being.

Subsistence economies

Subsistence economies are characterized by consumption of goods by the people or groups that produce them. Buying and selling is rather limited, and tertiary and quaternary activities are rare-to-unknown. Traditional societies that rely on manual labor and possess low levels of technological sophistication are typical of this type of economy. Four kinds of subsistence economies are commonly recognized and are discussed in the following sections.

Hunters and gatherers

Hunters and gatherers subsist on wildlife, fish, and/or parts of plants (roots, nuts, or berries) that grow wild. Contrary to a one-liner about a very different kind of work, this is unquestionably the world's oldest profession. Indeed, eons ago, before the discovery of agriculture, everybody did this. Today hunters and gatherers are few in number and are found mainly in isolated areas of tropical rainforest (Figure 15-4).

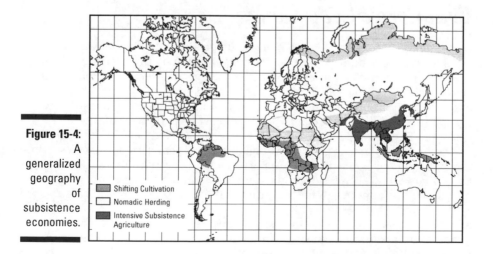

Figure 15-4:
A generalized geography of subsistence economies.

Shifting Cultivation
Nomadic Herding
Intensive Subsistence Agriculture

Nomadic herders

Nomadic herders tend domesticated herbivores (including sheep, goats, cattle, and reindeer) that feed on natural foliage during the course of a controlled itinerary (wandering). *Steppe* (grasslands) and tundra are the traditional realms of these people, sometimes called *pastoral nomads*. The global population of these folk is rapidly decreasing as farming occupies more land, and as people turn to *ranching* — the raising of livestock in fixed locales for sale instead of consumption. Also, governments in nomadic lands have engaged in campaigns to settle herders and bring them into the commercial (market) economy.

Shifting cultivation

Shifting cultivation involves farming on cleared plots of forest until soil fertility plays out; after which, a new plot is cleared and the old one abandoned. This agricultural method is described in greater detail in Chapter 10. Millions of people in tropical parts of Latin America, Africa, and Asia engage in this endeavor for their livelihoods. If anything, and unlike the previous categories, their numbers are increasing (probably substantially so) rather than decreasing.

Intensive subsistence agriculture

Intensive subsistence agriculture is characterized by high amounts of labor brought to bear on relatively small land holdings. Cultivation of rice paddies, which require considerable inputs of labor for construction and maintenance, plus precise applications of water at precise times, is perhaps the quintessential example. Well over a billion inhabitants of South and Southeast Asia engage in this type of subsistence activity.

Commercial economies

Commercial economies are characterized by production of goods and services for sale to others rather than for consumption by the producers themselves. Products do not enter into the distinction between subsistence and commercial economies and may, in fact, be the same. Rice, for example, is one of the principal foodstuffs that is grown and consumed by subsistence agriculturists. But commercial farmers, who eat little or none of their harvests, but instead sell it to companies whose boxes and bags of rice end up in your grocery store, also grow rice in quantity. As that example suggests, commercial economies may include primary economic activity, like agriculture. By and large, however, it's secondary, tertiary and quaternary activities (see "Categorizing Economic Activity" earlier in the chapter) that dominate.

In commercial economies, laws of supply and demand determine price and quantity. Competition largely influences decision-making concerning production, distribution, and — most importantly as far as this book is concerned — facilities locations. Indeed, one of the distinguishing features of commercial economies is deciding where an economic activity should be located.

Understanding Location Factors

A *location factor* is any aspect of an environment that influences a decision regarding the choosing of a site for business or industrial activity. Because competition is an essential characteristic of commercial economies, businesses of all sorts do things to give themselves a competitive advantage. Finding an optimal location for a business is a case in point, as well as an endeavor that has come to typify modern economic geography.

A good example of the importance of location concerns the decision to locate the plant that builds Saturn motor vehicles in Spring Hill, Tennessee, just south of Nashville. Theoretically, the factory could have been built just about anywhere. Indeed, after corporate officials announced their intent to manufacture "a new kind of car" in a brand-new plant somewhere in America, governors from several states lobbied to host the facility. And why not? The factory would create thousands of new jobs and provide a major boost to the economy of wherever it was built.

Saturn officials initially had no specific location in mind. Observers speculated, however, that the company's desire to build a high-quality product at the lowest possible price meant that several key characteristics would dictate the choice. Specifically, the chosen site would ideally have

- A large amount of inexpensive land on which to build
- Access to abundant and relatively inexpensive electricity
- A low tax environment compared to other prospective locations
- Connectivity to railway and interstate highway systems
- A central location with respect to the various subcontractors who would supply parts
- Centrality with respect to overall population distribution (which also would have a strong and direct correlation with the location of dealer's showrooms)
- An adequate supply of potential employees who would exhibit solid work habits for pay that was somewhat low by industry standards

A number of prospective sites in different states were considered. Most were strong candidates with respect to some of the above criteria, but weak in regard to others. In the end, Spring Hill got the nod because it satisfied all points.

Economic activities vary, of course, and as a result, so do the factors that are most appropriate in deciding where a particular economic activity may best be located. Thus, what is important with respect to deciding where to build cars may be totally different than the location keys to growing mushrooms or merchandising children's clothing. Also, in the case of some economic activities, a single location factor may be all-important, while in another, a mix of several determinants may need to be considered.

Obviously, a good business location may be the key ingredient to profitability or overall success. Accordingly, nowadays two of the most important functions of economic geography are to help enterprises identify the location factors that are best for their particular business, and to sift through the list of possible locations and choose a favorable site. The following sections offer a discussion of some of the more important location factors.

Proximity to raw material(s)

Proximity to raw materials may be a critical location factor for economic activities that involve manufacturing or processing. Both functions typically require delivery of raw materials from source areas to the place of business. The cost of transporting raw materials to where they are needed may be a major business expense, however. Thus, locating as closely as possible to the source or sources of those materials is a good idea to minimize production costs, as well as producing a cost-competitive final product. Following are three examples:

- **Cheese:** Cheese production consumes large quantities of milk, which is expensive to transport long distances because it can be bulky and requires refrigeration. Therefore, if you want to make it in the cheese business, you'd be wise to locate as close as possible to a bunch of dairy farmers. It comes as no surprise, therefore, that Wisconsin, a major dairy state, also leads the country in cheese production.

- **Steel:** The traditional blast furnace way of making steel requires lots of iron ore and lots of coal, but not in just any amount. Generally, production of 1 ton of steel requires about 1.5 to 2.0 tons of iron ore and 2 to 3 tons of coal. Both ingredients are very bulky and expensive to transport from mine to factory. But because more coal than iron ore is required, an old adage is adhered to that goes, "Steel follows coal" — meaning that steel plants tend to locate closer to coal fields than to iron ore mines in order to hold down transportation costs. Figure 15-5, a map of the traditional American "Manufacturing Belt" in relation to coal and iron, very clearly shows the relationship between steel and coal.

- **Hogs:** If you want to seek your fortune in pork or bacon, then you should locate your hog operation as close to corn country as possible. That is because corn is the principal ingredient in pig feed, which is something you are going to need by the ton. Like many domesticated animals destined for your refrigerator, hogs nowadays are raised in specialized facilities where they are fattened for market. Central to this process is bringing feed (often by the ton) to the pigs rather than setting the animals loose in the countryside to forage. Being near the feed source (corn) holds down transportation costs and contributes to profitability.

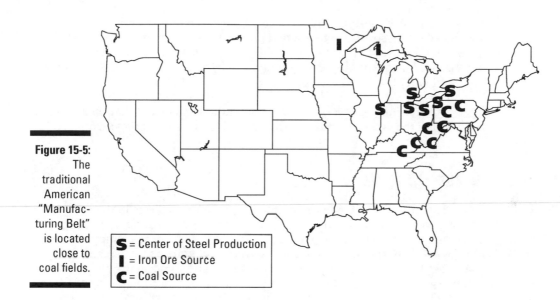

Figure 15-5:
The traditional American "Manufacturing Belt" is located close to coal fields.

S = Center of Steel Production
I = Iron Ore Source
C = Coal Source

Proximity to market(s)

Proximity to the market (the purchasers of products) for a business's goods or services may be a key location factor for some economic activities. The exact meaning of "market" varies from one enterprise to the next, however. If you're an ice cream vendor, then your market is people. If you're a zipper manufacturer, then your market might be companies that manufacture clothing. Here are three examples that illustrate the potential importance of proximity to market:

- **Pizza:** Pizza is a fairly low-cost item. Therefore, anybody who wants to make a profit in this business is going to have to sell a lot of it. An ideal location, therefore, is one that is close to a large number of people who are likely to want to have a fairly fast dining experience and not spend a whole lot of money. That could be a busy downtown area, a shopping mall, a freeway exit, a major highway intersection, or a site next to a college or hospital. (Hold the anchovies, please!)

- **Boutiques:** Boutiques typically specialize in high fashion designer wear intended for the upscale market. Unlike the pizza business, the main location strategy here is to get close to a particular variety of people (affluent folk) rather than lots of people in general. Locating in or near a high-income residential neighborhood or in a shopping district that caters to the well-to-do are good ideas.

✔ **Concrete:** Cities typically generate high demand for concrete, which is used for all sorts of construction projects. Once concrete is made at the plant and transferred to a concrete mixer, it needs to be delivered and poured at its intended construction site fairly quickly. Also, being a bulky and fairly low-cost item, concrete is not the sort of thing that can be transported profitably for a long distance. Thus, anyone going into the concrete business would be well-advised to locate his or her business within the city or on its edge as opposed to, say, out in the country.

Cost of labor

In the last couple of decades, numerous manufacturing firms in the United States and other developed countries have closed down and relocated to developing countries where people do the same work as, say, American counterparts, but for much less pay and fewer (if any) benefits. As a result, lots of manufactured goods that used to have "Made in the USA" emblazoned on them now say "Made in Malaysia," or the Philippines, or China or somewhere else. Believe me, if you aren't aware of this, then you have a major geography lesson waiting for you in your clothes closet. Check out the labels on your clothing and shoes. Chances are good that many, if not a majority, are imports. The two following sections give examples.

My sports jacket

Consider the sports coat I wore to work today. It has two labels inside. One says "100% camel's hair" and the other says "Made in Costa Rica of USA components." That means the company that makes the jacket imports camel hair cloth, probably from another company in the Middle East, and then cuts it up into components that will be sewn together to make apparel in different sizes — a "40 Regular" in my case.

But instead of having the sewing done in the USA, the manufacturer ships it off to Central America. There, seamstresses perform their work probably for a third or less of what the company would have to pay an American worker. Afterwards, the jackets are sent back to the U.S. for sale. Even though it costs a bundle to send the bundles back and forth, the money saved by having the sewing done by comparatively inexpensive foreign labor more than makes up for the transportation costs.

Data processing

Manufacturing is not the only activity that has gone abroad to take advantage of low-cost labor — so has data processing. For example, a well-known American insurance company receives premium payments at a U.S. address, but then ships everything to Ireland for processing. Another company does basically the same thing, but has its data processing done in Ghana, West Africa. Similarly, a well-known American airline has some of its data and time-card processing done in Barbados. In each case, the prime rationale is to save money on labor costs.

TECHNICAL STUFF

Atlantic City as intervening opportunity

An *intervening opportunity* is a closer source of supply to a point of demand. Clear as mud, right? Okay, suppose you occasionally need to purchase widgets, and the closest store that sells them is 10 miles away. Then one day another store that sells widgets opens up 2 miles away. From your perspective, the second store is an intervening opportunity because it's a closer source of supply (widgets) to a point of demand (you). And because of that, you decide to discontinue traveling to the first store and become a patron of the second one. Naturally, the proprietors of prospective new stores have a keen interest in locations that offer competitive advantage. And for that reason, entrepreneurs often scope out the competition and then choose a location that represents an intervening opportunity for as many potential customers as possible.

A good real-world example of intervening opportunity concerns the advent of gambling in Atlantic City, New Jersey. Back in the early 1970s, gambling within the United States was legal only in the state of Nevada. Basically, it was a monopoly from which Las Vegas and

Reno profited handsomely. But then a plan was hatched by New Jersey state legislators to permit casino gambling in Atlantic City, an old seaside resort that was a bit down in the dumps. The promoters clearly saw that Atlantic City would be an intervening opportunity for millions of Easterners for whom a few hours' drive to Atlantic City would be preferable to a trip of nearly transcontinental proportions. Indeed, when the casinos finally opened, business boomed. At least for a while.

Gambling was so successful in Atlantic City, that it gave people in other states the idea of doing the same thing. In particular, people on Indian reservations were attracted to the idea because it promised income and because Native Americans did not need the approval of state legislators and voters to open casinos. (Reservations are technically not state land.) So casinos started popping up on reservations all over America, including the Northeast. And as they waxed, Atlantic City waned because it no longer enjoyed its status as the singular intervening opportunity for eastern gamblers.

Accessibility

Accessibility concerns the ease with which a business can be reached, or with which goods can be transferred from one location to another. A downtown shopping area, for example, may be close to a lot of people, yet not be very accessible because of traffic congestion and lack of parking. Thus, ease of transportation is key to facilitating the accessibility of a particular location. The following sections give some examples of how businesses use accessibility to their benefit.

Hub-and-spokes networks

Nowadays, virtually every airline employs a hub-and-spokes network in which passengers or freight fly from different locations to a central point (*hub*), where transfer is made to another plane to reach the final destination (*spoke*).

While this inevitably involves more travel time versus a non-stop direct flight between the points of origin and final destination, it also affords accessibility between all points in the route system and income maximization to the carrier.

Federal Express, for example, the company that pioneered the overnight package delivery business, operates a hub in Memphis, Tennessee (Figure 15-6). In the middle of the night, company aircraft from all over America bring their packages to Memphis, where they are swapped between planes that then return to their points of origin in time for morning deliveries. Thus, each night a company airplane departs from, say, New Orleans with packages destined for cities all over the country. But instead of delivering the goods "Santa Claus-style," going to each city one by one, the aircraft goes to just one place — Memphis, Tennessee.

Aircraft from dozens of other cities do the same thing, dropping off packages destined for New Orleans and elsewhere, and then picking up packages from New Orleans and elsewhere before making the flight back home. Thus, while it may seem a bit weird that a package sent from New Orleans to Phoenix makes the trip via Memphis, the network design promotes accessibility between multiple locations using a minimum of aircraft that fly a minimum number of segments. And that promotes profitability.

Maquiladoras

In recent years, numerous American manufacturers have engaged in *outsourcing* — producing components or products abroad for domestic sale — to take advantage of lower-cost foreign labor. This tactic is exemplified by *maquiladoras,* American-owned manufacturing plants located on the Mexican side of the U.S.-Mexico border. Cross into Ciudad Juarez, Tijuana, Nuevo Laredo, and other Mexican border towns and you are likely to find plants bearing familiar U.S. brand names producing clothing, toys, leather goods, electronic items, auto parts, and other goods destined for sale in the U.S. Their presence is due in large measure to Mexican laws that allow U.S. companies to import components duty-free for assembly in factories located within 20 kilometers (12 miles) of the border. Once completed, materials are sent back across the border for final assembly or distribution to retail outlets. Thus, the arrangement allows U.S. manufacturers to profit by lowering their production costs and Mexico to profit from job creation and (in some cases) technical training. One might characterize this as an unequivocal win-win situation, except that opening a *maquiladora* is often complemented by closure of a manufacturing facility in the U.S., resulting in people out of work.

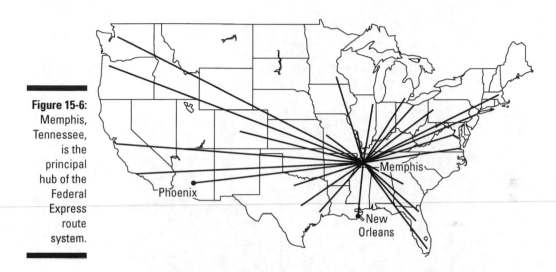

Figure 15-6:
Memphis, Tennessee, is the principal hub of the Federal Express route system.

Shopping malls

The success of a mall rests on its accessibility to a large number of potential shoppers. For that reason, nearly every shopping mall you'll ever visit is located beside a major multi-lane highway (and perhaps by a busy intersection as well) that accommodates a considerable volume and flow of traffic. Of course, all of those potential shoppers will need to leave their cars somewhere, so the shopping complex is always adjoined by a several-acre parking lot — which in itself is a key component of accessibility. Altogether, a substantial amount of land is required. That argues for a location on the urban fringe, where land in quantity is most likely to be available. In addition, in most American cities the people who have the most disposable income are more likely to live closer to the urban fringe than to downtown. Thus, an additional "fringe benefit" is proximity (which in this case equals accessibility) to the very people that mall operators most want to attract. In the case of major suburban malls, the quintessential location is next to an intersection by a beltway.

Containerization

As suggested in the discussion of labor costs, a company, such as an American athletic shoe manufacturer, may operate a factory in a foreign country in order to take advantage of low-cost labor. But if the finished shoes are not accessible to the American market — that is, if they cannot then be shipped to the United States cheaply and efficiently — then the whole purpose of manufacturing the product overseas is rendered pointless.

Figure 15-7 illustrates *containerization,* the shipping of goods in large metallic containers that can be loaded and off-loaded from ships directly onto tractor-trailer trucks for onward transportation to the final destination. This technology has revolutionized maritime trade by

 ✔ Reducing loading and unloading time of ships from days to hours

 ✔ Facilitating prompt delivery of goods

 ✔ Reducing pilferage and breakage that occurred when goods were loaded and unloaded bit by bit from ship's holds

Figure 15-7:
A container
ship unloads
at port.

Special port facilities are required for the loading and unloading of containers. For that athletic shoe manufacturer, therefore, finding a country that offers cheap labor may not be enough. It may also have to have a containerization port facility to make foreign manufacturing sufficiently profitable.

Being complementary

Complementarity is the tendency for companies that produce goods and services for each other to cluster together, thereby maximizing mutual accessibility. New York City's Theater District, for example, consists of more than just venues for Broadway shows. Companies that act as agents for performers, produce sets, sell lighting and sound equipment, make costumes, coordinate ticket sales, and rent out sound stages and rehearsal studios can also be found on Broadway. You don't have to travel to New York to witness complementarity, however. It's a basic principle of economic geography. So if you live in a fair-sized city or town, chances are good you can find one or more examples of associated businesses that have clustered together for mutual benefit. For example, if your town contains a fair-sized hospital, then doctors' offices, insurance representatives, and medical laboratories are likely to be located close-by.

Cost of land/rent

Companies require land or rental space where they can conduct business. Within any sizeable geographical area, however, the cost of land or rent varies. Locating a business (or part of a business) with respect to this variation may have a significant impact on the company's bottom line.

Take, for example, a well-known bank that is headquartered in New York City. When you mail them your credit card bill, the letter goes to an address in South Dakota. Another well-known New York City department store receives and processes its credit card payments at an address in the state of Georgia. To some extent, geographic differences in the cost of labor are at work here. But in these cases, another and more important location factor is at play: geographic differences in the cost of land and rent. Simply put, these operations require lots of office space, the price of which is much cheaper in towns in South Dakota and Georgia than in New York City.

I'm going to temper my discussion of this location factor because it appears again in the chapter on urban geography (see Chapter 17). Suffice it to say here that supply and demand largely determines the cost of land/rent; and that intense demand and competitive bidding for land/rent is greater in big cities than in small ones, leading to higher land costs in major urban areas. Indeed, land/rent prices tend to "spike" downtown, and then trail off dramatically with increasing distance from urban cores, as evidenced by Figure 15-8.

Figure 15-8:
A generalized urban rent gradient.

Figure 15-8 has major implications for business locations. Activities that locate downtown must be capable of covering operating costs that include high rents. If they cannot, then they are wise to locate at some distance from downtown, where land/rent costs are likely to be much less. Or perhaps even to a town in South Dakota or Georgia.

Taxes

They say that the only sure things in life are death and taxes. Maybe so. But taxes, at least, can be minimized. That is because tax laws and tax rates vary geographically. Thus, if the tax laws in a particular locale are so high as to seem unfavorable to personal or business finances, then one may consider moving to a more favorable location.

Local variation

Tax structures differ from one country to the next and, within the United States, from one state to the next (see Figure 15-9). Some states have no income tax, while in those that do, the rate schedule varies. The same is true of corporate income tax and state property tax — some states have them and some don't.

On the other hand, every state has a sales tax, but rates vary substantially from one state to the next. Also, particular items may be exempt from sales tax in particular locales. In New York City, for example, the standard sales tax is 8.25 percent, but clothing is tax-exempt. The purpose of that is to encourage New Yorkers to shop at home and support local businesses rather than cross the Hudson River to purchase goods in the comparatively low-tax environment of New Jersey.

Enterprise zones

As the foregoing vignette suggests, "creative taxation" may be used to encourage business activity in particular locales. One fairly popular option is to designate corporate tax-free areas that encourage job creation in particular locales. Called *enterprise zones* (or *empowerment zones*), the typical goal is to counter unemployment and stimulate the economy of an area that has fallen on hard times.

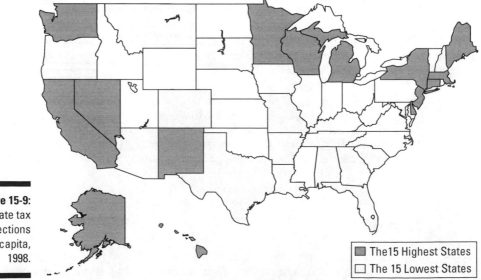

Figure 15-9:
State tax collections per capita, 1998.

■ The 15 Highest States
☐ The 15 Lowest States

Applied Geography: "Start spreading the news"

An advertisement once appeared in a business magazine that showed the Manhattan skyline and the headline, "Start spreading the news: Alabama is open for business." The accompanying text encouraged executives in New York City to consider relocating their businesses to Alabama. Compared to New York City, the ad pointed out, Alabama has much lower taxes, land costs, housing prices, and utility rates. Thus, considerable advantages could be realized by moving one's business from Manhattan to, say, Mobile, Alabama. It also inferred that workers could be paid less, but still come out ahead. Suppose, for example, a certain employee in New York City receives an annual salary of $50,000. In Alabama you could pay somebody, say, $45,000 (or less) to do the same job. But because of the tax and other differentials mentioned above, the Alabama employee could well end up with more disposable income in his or her pocket than the counterpart in New York City.

To use a hypothetical example, suppose a certain city with an old industrial core has witnessed several plant closings, resulting in derelict real estate and an increase in unemployment. One turn-around possibility is to designate all or part of that area as an enterprise zone and mandate that any company that employs, say, at least 50 people and locates within the area will be exempt from corporate taxes. Theoretically, at least, this is a win-win-win situation. The company gets a tax break that may increase its bottom line, unemployed people get jobs, and some of that income is returned to the government in the form of sales and income taxes.

Climate

Climate is an obvious location factor for various agricultural and leisure pursuits. Thus, citrus fruits are produced in extreme southern (warm) parts of the U.S. because the plants are very sensitive to frost. Wheat, in contrast, is a hardy grass that does not require much moisture. Much of it is grown, therefore, in low-to-modest precipitation environments of the U.S., Canada, and Australia. Regarding leisure, ski resorts obviously call for a certain kind of climate as do most modern beach resorts. Other businesses, however, have climate connections that are a bit more subtle, but no less important. Following are two general examples.

Textile manufacturing

When raw cotton or wool is spun into thread, there is a propensity for the fiber to snap. This likelihood is particularly high in dry climates and comparatively low in humid climates. Thus, climates characterized by moderate-to-high atmospheric humidity are superb for textile manufacturing. Britain and the Southeastern U.S. are in this category. And both places are characterized by textile manufacturing.

A matter of amenity

In the case of quaternary economic activities (and many tertiary ones, too), good personnel are critical to business success. If, therefore, a company can offer employees opportunity to live and work in a region that is generally perceived to have an attractive physical environment, then it may enjoy a competitive business advantage over competing firms. Thus, many high tech companies have chosen their locations on the basis of a balmy climate (as much as any other location factor), in the hope of attracting talented employees.

Looking Toward Location Trends of Tomorrow

Every year more and more people gain Internet access and make purchases online. Indeed, several well-known "dot.coms" conduct business only in cyberspace as opposed to the real space of a commercial district or mall. Because these companies are sometimes hailed as a prelude of things to come, and because they have liberated their customers from having to travel to a place of business, some people have suggested that in the future, location factors will not be as important as they are today.

I beg to differ. Many (perhaps most) kinds of business will continue to operate best in the real-world environment of face-to-face contact. And, of course, businesses in cyberspace actually exist someplace in the real world. That is, somewhere they have real, live employees who sit at desks and maintain the Web site, process purchase requests, maintain a warehouse or warehouses, do the actual packaging and mailing of merchandise, and handle returns and refunds. While the Internet may afford companies a certain freedom of location, they will nevertheless end up somewhere on planet Earth. And as always, some locations will prove to be more economically advantageous than others.

Chapter 16

An Appetite for Resources

· ·

In This Chapter

▶ Changing tastes and preferences

▶ Making a culture — resource connection

▶ Deciphering the geography of supply and demand

▶ Looking at the advantages and disadvantages of different resources

· ·

*P*eople have an appetite for resources; and as their numbers grow, so does the cumulative thirst for Earth's bounty. The global cornucopia is varied, however. Resources differ in their abundance and longevity — meaning that some exist in finite supply while others are eternal. Geographical considerations further complicate the picture. People in different parts of the world have different appetites for different resources that typically are available in some regions but not others.

Supply and demand for Earth's resources is a geographically complex and fascinating story. The goal of this chapter is to touch upon and familiarize you with some of the key concepts of resource geography.

Energy resources serve as the focus of this chapter. This somewhat narrow perspective is certainly not intended to denigrate the importance of other kinds of resources, all of which lend character to different parts of the planet. Indeed, discussions of soil, fisheries, and forests appear in Chapters 7, 8, and 10 respectively. Likewise, much of Chapter 18 is devoted to consequences of resource utilization. For now, the key thing to remember is that *resource* entails a great many things that range from specific environmental items to the land itself. Energy resources, however, are of critical importance the world over, and embody the key concepts and challenges common to most resources.

Defining Resources and Assessing Their Importance

A *natural resource* is any physical environmental item that people perceive to be useful for their well-being. That's a pretty broad definition, but an appropriate one. All kinds of environmental items are useful. Geographically, resources may be everywhere (air), pretty widespread (water), limited to relatively few areas (petroleum), or downright rare (diamonds). As the following sections will demonstrate, however, the geography of resources, as well as the wealth and power they provide, may have more to do with culture than the environmental items themselves.

The central role of culture

An energy crisis is currently underway. Fuel resources are becoming scarce, and demand for them is growing faster than new sources are becoming available. Prices are going up. People wish for self-sufficiency, but that is no longer possible. Increasingly, they must rely on fuel sources that are far away, under the control of somebody else, and require increasing amounts of time and money to procure. No immediate solution is in sight.

No, this is not about petroleum or the United States, but instead the firewood crisis in parts of sub-Saharan Africa. In some countries on that continent more than 75 percent of all energy consumption involves burning firewood, mainly for cooking and boiling water. Population is growing and with it the demand for firewood. But practically nowhere in the steppe and savanna realms, where scarcity of firewood is most acute, are there effective programs that plant or replant trees faster than they are cut. Woodlands that used to surround a village are gone, and therefore people must walk farther and farther to get what they need.

Resources are culturally determined. That is, whether or not a particular environmental item is useful depends on culture. Many Americans could care less about access to firewood, but they certainly do care about access to gasoline. Lots of Africans feel the exact opposite. Cultural difference, particularly as it relates to technology, is the principal explanation. American culture is inseparable from a wide range of machinery that runs on gasoline or oil — cars for transportation, tractors to help produce food, furnaces that heat buildings and homes, and so forth. Over large parts of Africa, however, none of these matter very much, but firewood sure does. The bottom line is that American culture is different than African culture, which are different than Asian culture, different than Eastern European culture, and so on and so on. What this means is that people from different cultures use and evaluate the environment differently.

Culture change, resource change

All cultures change over time. As they do, certain items of everyday use (horse-drawn buggies) may gradually disappear, while previously unknown items rise in importance (automobiles). Because resources are culturally defined, it stands to reason that culture change may result in resource change. Consider, for example, changes in energy resource consumption that have occurred in the United States (see Figure 16-1).

Figure 16-1:
Sources of
energy in
the United
States: 1850,
1910, and
1990.

In 1850, 90 percent of the United States' energy supply consisted of firewood. Virtually all of the remaining 10 percent came from coal. By 1910, however, those numbers were virtually reversed. About 78 percent of the country's energy consumption was based on coal and about 7 percent on firewood. The remaining 15 percent came from formerly untapped or unused sources: oil (6 percent), natural gas (6 percent), and hydropower (3 percent).

By 1990, the picture had completely changed once again. Oil was now the most important commodity, accounting for about 39 percent of all energy consumption. The percent contribution of coal had shrunk to about 24 percent, while that of natural gas had risen to about 25 percent. The contribution of hydroelectric remained a fairly steady 4 percent. Finally, nuclear energy had come on the scene and was contributing about 8 percent to the national total while an old stalwart, firewood, had virtually dropped off the charts (though it remains important in home heating in a few states.)

These changes had nothing to do with depletion and discovery of resources, but a great deal to do with culture change. Thus, decreased use of wood over time is not explained by depletion of forests. In fact, more forest cover exists today in the contiguous 48 states than there has been for some time. Likewise, lack of reliance on oil and natural gas 150 years ago was not due to lack of these items or knowledge of their existence. Instead, there were limited means of putting them to productive use, and, therefore, they were in very limited demand.

But times change. Which is to say culture changed. And with it came a change in energy resource consumption as machinery was developed that made use of oil and gas instead of wood.

Resources and power

Fuel resources generate two kinds of power. On the one hand, they can be used to make heat that can be transformed into physical energy. In addition, resources may generate significant political and economic power for the countries that possess them.

Petroleum, for example, comes as close as any substance today to being the economic lifeblood of the global economy. Its geography is highly concentrated, however (as seen in Figure 16-2). The countries that border the Persian Gulf possess about half of all the world's known petroleum reserves. Saudi Arabia alone possesses nearly 23 percent. Elsewhere, major reserves exist in Russia (13 percent of the world's total), Venezuela (6 percent), and Mexico (4 percent). The United States possesses thousands of producing oil fields. Most are small, however, so overall it isn't a leading center of reserves.

The essential fact is that few countries possess petroleum in great quantity. That virtually guarantees that those countries will enjoy substantial political and economic clout for as long as petroleum maintains its status as the most important energy source. This was vividly demonstrated in 1973 when, in the wake of an Arab-Israeli war, oil-producing nations in that region curtailed shipments to western countries. About 60 percent of the world's petroleum reserves then came from countries that border the Persian Gulf. Much of it was exported to developed nations whose economies depended on it to different degrees. When the tap suddenly was turned off, everybody got a quick and unmistakable lesson in political petrol-power.

But that kind of power is obtained only when a particular resource is sorely needed by countries that either have none or not enough. We have seen that resources are culturally determined, and that culture change brings resource change. Accordingly, as resources rise and fall in importance, so do the economic and political power of the places that possess them. One hundred years ago, the countries that border the Persian Gulf were of little economic importance to the outside world. Times have changed. Today, virtually every country consumes petroleum, and the vast majority of them cannot satisfy their thirst by domestic production. So they rely for their economic lifeblood on foreign sources — the Persian Gulf nations in particular — who accordingly assume great power.

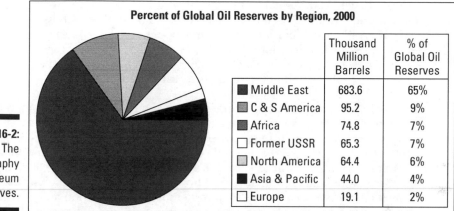

Percent of Global Oil Reserves by Region, 2000

	Thousand Million Barrels	% of Global Oil Reserves
■ Middle East	683.6	65%
▨ C & S America	95.2	9%
■ Africa	74.8	7%
☐ Former USSR	65.3	7%
▨ North America	64.4	6%
■ Asia & Pacific	44.0	4%
☐ Europe	19.1	2%

Figure 16-2: The geography of petroleum reserves.

Resources and wealth

You may expect countries that possess lots of natural resources to be better off than those that don't. For better or worse, however, it doesn't always turn out that way by a long shot. In fact, some poor countries are rich in natural resources while some rich countries are poor in natural resources. This difference may occur for a number of reasons, perhaps the most important of which concerns the locations of people and facilities involved in the acquisition and processing of raw materials.

Turning to another resource for a moment, consider the geographical differences in employment and income that may occur when trees in "Country A" are cut down and sent to "Country B" for processing. Several people may be involved in harvesting the forest, from which timber is sent to a sawmill in "Country B." Many more people may be employed at a sawmill, where raw logs are cut into pieces of lumber, than were employed back in the forest of "Country A" to cut the trees. Also, because the factory employees are skilled workers, they are likely to command higher salaries as well. Later, skilled craftspersons might take pieces of that lumber and make high-priced furniture. Finally, company executives negotiate sales and contracts with transportation firms, wholesalers and retailers concerning the final disposition of finished products.

Thus, a single log can generate income for several people. But the amount of income and the location of the beneficiaries varies. Generally, the person who cuts down the tree probably earns less than the sawmill worker, and so forth. Therefore, in this example, the exported log generates much more wealth where it is processed and made into finished products than where it was harvested.

Under these hypothetical conditions, therefore, "Country A" would realize comparatively little income or wealth from its natural resources, while "Country B" (which perhaps is resource-poor) would realize a great deal. Incidentally, a law in Indonesia prohibits exportation of raw logs. What may sound like a silly edict is in fact an effective means of helping that country to realize the full income potential of its forest resources.

Though this example concerns trees, other natural resources — mineral ores and crude oil, for example — may exhibit similar outcomes. Formerly, Persian Gulf oil was drilled by foreign-owned companies that set the prices, employed relatively few locals, and sent crude oil home for processing and refining, resulting in a finished product worth substantially more than what came out of the ground in the first place. Now, however, the producing countries have a much greater stake (if not outright ownership) in production and processing, and are realizing close to the full economic potential of their resource base.

Differing Life Spans: Which Resources Are Here Today or Gone Tomorrow

Resources have different life spans. Some exist in finite quantity; so when they are used up, they're gone forever. Others can be replenished, as when seedlings are planted to replace a forest that was cut down. And some are virtually eternal (the sun, for example), meaning they'll always be with us regardless of how humans use the environment. Consideration of energy resources that fall into the following categories provides insight into resource geography, the crux of the current energy situation, and the need for future planning.

Non-renewable resources

Non-renewable resources exist in *finite* (or fixed) quantity. You can think of them as coins in a global piggy bank from which money can only be extracted, not added. Once they have been used up, they are gone and cannot be renewed.

Perhaps the most important energy-related fact of life is that the United States and other developed countries are overwhelmingly dependent on non-renewable fuel sources. That includes petroleum, coal, and natural gas. These are sometimes called *fossil fuels* inasmuch as scientists believe that they are the result of long-term decay and metamorphosis of organic matter. Thus, even as I write this, nature is at work doing whatever it does to turn, say, today's ocean bed into tomorrow's oil field.

But that process takes millions of years, in contrast to humanity's needs, which are exhausting the world's oil inventory in a figurative blink of the eye.

The world's first commercial oil was drilled in Pennsylvania in 1859. Now, a mere century-and-a-half later, people are contemplating the end of the "petroleum era." But I'm getting a little ahead of the story.

Petroleum

Next to air and water, petroleum is perhaps the most essential resource of the moment, at least as far as developed countries are concerned. Arguably it can be considered the lifeblood of the American economy, if not the American way of life. Skeptics need only imagine waking up one morning to discover that every car, truck, motorcycle, and internal combustion engine no longer functions for lack of gasoline.

Presently, global petroleum reserves stand at about 158 billion barrels (see Figure 16-2 for a look at the geography of petroleum), and are being consumed at a rate of 3.1 billion barrels per year. That means we currently have about a (158 ÷ 3.1 =) 51-year supply of oil.

In the years ahead, new petroleum deposits are likely to be discovered, but global demand for oil is also likely to rise. How this math will play out can't be accurately determined at present, but clearly the days of the "petroleum era" are numbered. Obviously, if a replacement for petroleum is not found, its price will rise even further as reserves decline. As that occurs, the economic and political power of countries that possess it will continue to increase. On the other hand, growing scarcity is also likely to increase development of alternative sources of energy.

Coal

The world's coal supply greatly surpasses petroleum in quantity and longevity. The estimated global coal reserve is in excess of 1 trillion metric tons while annual production is about 4.6 billion metric tons. That works out to nearly a 220-year global coal supply at the current rate of consumption. The United States possesses nearly 25 percent of all of the world's known deposits, and at the current rate of production (mining) that will last for approximately 250 years.

If you're thinking, "Gee, that's terrific," then you need to temper your optimism for two reasons. First, most coal burns dirty, so its use comes at the expense of air quality (see the "Grades of coal" sidebar later in the chapter). As a result, substantial research dollars are being spent to find ways to burn coal without polluting the air — such as by removing impurities before being burned or by developing smokestacks that are akin to giant filter cigarettes.

Second, coal is bulky and unwieldy. Although it can be crushed, mixed with water, and transported short distances by pipeline as slurry, you can't send coal down a pipeline a long distance or pour it into a gas tank — although it would be a much more convenient and widely applicable if you could. Thus, considerable research has focused on coal *liquefaction*.

In any event, coal, like petroleum, is very unevenly spread across the planet (as shown in Figure 16-3). While substantial reserves are found in North America, Europe, and Asia, South America and Africa are largely without any. Thus it has limited geographic prospects. In addition, and as discussed in the sidebar, some coal reserves are much more valuable and useful than others.

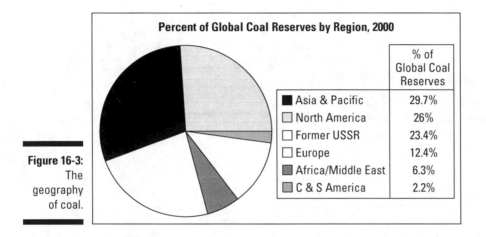

Figure 16-3:
The geography of coal.

Percent of Global Coal Reserves by Region, 2000		
	% of Global Coal Reserves
■ Asia & Pacific	29.7%
☐ North America	26%
☐ Former USSR	23.4%
☐ Europe	12.4%
▨ Africa/Middle East	6.3%
▨ C & S America	2.2%

Natural gas

Natural gas exists in abundance and, as far as fossil fuels go, has attractive qualities. It burns very cleanly (producing only the most modest air pollution), requires little or no processing before use, and can be transported cheaply and efficiently overland by pipeline. Its versatility is also a point in its favor. In addition to heating a majority of the homes in the United States, natural gas is used to generate electricity, provide energy for some kinds of light manufacturing, and power an increasing number of motor vehicles.

On the downside, natural gas is not well-suited for transoceanic trade. Pipelines of that length are not feasible. Shipping is possible when *liquefied natural gas* is involved, but that is not cost effective. Also, natural gas is, after all, a non-renewable resource, so someday supplies will be exhausted.

That is not going to happen anytime soon, however. Given current data on reserves and annual production, the world has about a 70-year supply of natural gas. But people who know the pertinent geology all suggest that vast quantities are waiting to be tapped in various parts of the world. The potential reserves of Alaska alone, for example, are thought to be double that of the rest of the United States.

Grades of coal

Coal exists in several varieties, or *grades,* the difference being determined by the degree of carbon content. *Anthracite*, the highest grade, is almost pure carbon and therefore burns very hot and gives off little smoke or soot. *Bituminous,* the next highest grade, has somewhat less carbon content and a higher percentage of waste materials. As a result, it burns less hot (but still hot enough for most industrial purposes) but dirtier, generating more pollution. *Lignite* and *peat* are progressively lower grades, burning even less hotly, while generating more air pollution.

In nature, sadly, the most pure form of any substance tends to occur in the smallest quantity. Thus, all global coal reserves considered, anthracite accounts for the smallest portion. Being so pure and burning so hot, it is also the most sought after variety. The end result is that that particular kind of coal is mostly depleted. Thus, the vast majority of the world's remaining supply consists of "dirty varieties" whose consumption poses challenges to environmental quality.

A global map of natural gas reserves, therefore, must be presented with the warning that significant changes will be seen in a matter of decades (see Figure 16-4). For now, Russia is home to about 34 percent of all global natural gas reserves. Iran possesses another 15 percent. The United States possesses only about 3 percent of global reserves, but accounts for about 23 percent of annual worldwide production. Thus, the U.S. consumes its reserves at a rapid rate. As noted, however, the prospects for new discoveries are good.

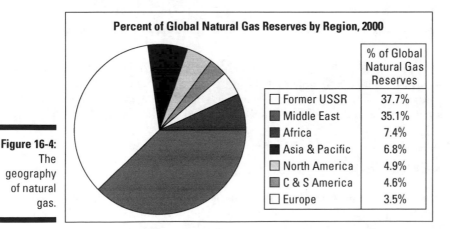

Figure 16-4: The geography of natural gas.

Percent of Global Natural Gas Reserves by Region, 2000

	% of Global Natural Gas Reserves
☐ Former USSR	37.7%
■ Middle East	35.1%
■ Africa	7.4%
■ Asia & Pacific	6.8%
☐ North America	4.9%
■ C & S America	4.6%
☐ Europe	3.5%

Applied Geography: The nuclear dilemma

The "dilemma" concerns three questions. Should more nuclear power plants be built? If so, then where? And where should we store the nuclear waste that we need to deal with in any event?

Nuclear power is a technologically sophisticated means of producing electricity. Specifically, a nuclear reaction induces rods of enriched uranium to produce high heat but also toxic radiation. This occurs in a heavily leaded nuclear reactor chamber that captures and contains the radiation even as it becomes very hot. To keep the chamber at a tolerable temperature, water must continually swirl around it. Steam is a by-product; and under pressure it may be used to turn a turbine (propeller) that operates a generator that produces electricity.

The United States is basically self-sufficient in uranium. Thus, proponents of nuclear energy see it as an important and under-utilized alternative to fossil fuels, and a way of diminishing our reliance on foreign sources of energy. The general public, however, is deeply concerned about the safety of these facilities (due in good measure to the Chernobyl disaster), so future construction is far from certain.

Two factors have historically guided the location of nuclear power plants: immediate access to a reliable water supply and proximity to users. Continuous water input is needed to guarantee continuous cooling of reactor chambers and with it continuous production of steam. Facilities therefore have tended to be located in coastal environments or along large rivers and lakes that have a history of not flooding. Proximity to users helps keep down costs of delivery and also conserve the electricity that has been produced. Not surprising, however, most people are reluctant to live close to one of these facilities. The dilemma, therefore, is trying to find locations that satisfy both the necessary water and the economics of proximity to large numbers of users.

The waste dilemma is perhaps even more daunting. After a certain period of use, the uranium rods lose their capacity to produce heat, yet continue to generate radiation at levels that may remain deadly for centuries. Finding a location or locations where this waste can be safely stored for a prolonged period of time is key.

However the map may change, the benefits of natural gas — like fossil fuels in general — will be unevenly distributed. Some areas are certain to have lots of it and others none. But as far as non-renewable resources are concerned, the future of natural gas probably burns brightest.

Renewable resources

Renewable resources are ones that can be replenished and, therefore, are theoretically inexhaustible. As mentioned at the beginning of this chapter, trees are an example because they can be replanted. Another is *biomass conversion,* which is the processing of organic matter into combustible liquids or gases. These forms of energy are "renewable" because the organic matter used to produce them can be grown (replenished) on a continuous basis.

In Brazil, a country that is poor in fossil fuels, sugar cane is used to make alcohol that is mixed with petroleum to form *gasohol*. Millions of vehicles in that country run on this substance, which is only about 20 percent gasoline. Elsewhere, organic wastes are locally collected and fermented in a low-tech way that produces a methane *biogas* (gas) used for heating, cooking, and lighting. China and India, for example, have extensive national biogas programs.

Of particular relevance to this book is the fact that biomass conversion may potentially break the bonds of location. Fossil fuels exist in limited, specific locales. In contrast, the number and extent of regions that can and could grow, say, sugar cane or corn, in abundance for purposes of biomass conversion is robust. The Corn Belt across the United States Midwest, for example, may one day conceivably become the Fuel Belt, helping to break the dependence of a fuel-hungry world on those relatively few locations where fossil fuels are found.

Perennial resources

Perennial resources are theoretically eternal. While they offer great promise, all have current limitations that relate directly to geography. Here are four examples.

Solar energy

Although astronomers tell us the sun will eventually burn out, for all intents and purposes, it is an eternal, perennial source of power. Energy from the sun can be absorbed directly by solar panels and put to use for heating. Even better, however, are *photovoltaic cells,* which convert sunshine to an electrical current. The sun is fickle, however. Some days it shines and some days it does not, so a major engineering problem is how to store solar electricity from sunny days to carry us through cloudy days. Also, knowledge of climates informs us that solar intensity varies around the world (see Chapter 9).

Unfortunately, areas of maximum energy need often occur where the sun is comparatively weak (or often obscured) and vice versa. Thus, solar and other energy sources that possess a measure of unreliability probably should not be looked to as complete solutions to future energy needs. Better perhaps that we view them as parts of a resource menu from which different resources can be called upon as conditions permit.

Wind energy

Wind is physical energy that can be harnessed to produce electricity. This is achieved by constructing rather specialized windmills that are tall poles (or stanchions) fitted with a propeller (actually a turbine) at the top. When the wind blows, the propeller turns. This rotating motion is linked to and operates a generator that produces an electrical current.

Wind is a by-product of sunshine (see Chapter 9); and because solar energy is a perennial source of energy, wind power is categorized as the same. But like the sun, wind is fickle. Some days it blows and some days it doesn't. Accordingly, for wind to become a major contributor to energy production, means must be found to store surplus wind-energy from windy days for consumption on calm days. Also, some places are inherently windier than others. Coastal areas, for example, tend to be particularly breezy because water and land absorb sunshine at different rates. Mountain and foothills areas also are especially favorable.

Given present technology, a single windmill provides sufficient energy for only a limited number of consumers. Hundreds would be needed to provide for the needs of a small city. And indeed, in some locales substantial real estate has been devoted to *wind farms* that are testing and applying large-scale feasibility of this energy type. Regardless of their success, these "farms" clearly represent a new addition to the cultural landscape (Figure 16-5).

Figure 16-5:
Wind farms such as this one may become more common-place as fossil fuels become depleted.

Geothermal energy

In the opinion of some, we are virtually standing on top of all the energy we will ever need. That is a reference to the very high temperatures that exist within the Earth's crust (see Chapter 6). Because this heat literally radiates from the Earth, scientists refer to it as *geothermal energy*.

Theoretically, all that is needed is to drill down to the level of very hot rock, inject water, and extract the resulting steam. This may be used directly to provide home heating and hot water. Under pressure, steam is also a physical force (witness the tiny whistling steam jet on your tea kettle, only imagine a huge one) that can be used to turn a turbine (propeller) that operates an electrical generator.

Geothermal energy is free and non-polluting. But a major impediment stands in the way of its widespread use — cost. In most places suitably hot rock is encountered about 2 to 3 miles beneath Earth's surface. Drilling and operating steam-producing wells that deep would not be cost competitive with standard coal-fired plants. Some locations are exceptions, however. In lands near plate boundaries, the attendant fissures make it possible for interior heat to come close enough to the surface to allow for economical human exploitation. In Iceland, for example, nearly all home heating and hot water comes from geothermal energy. In parts of Northern California (and a few other locations worldwide), electricity is produced in this manner. In the future, geothermal energy may become generally available as a result of technological progress and the changing economics of energy production. For now however, its viability is governed largely by the geography of plate boundaries.

Hydroelectric power

Hydroelectric power (HEP) utilizes the movement of water on Earth's surface to produce electricity. It is most often associated with a dam that is built across a fairly narrow valley, causing a river that runs through it to back up on the upstream side. Eventually, a reservoir rises to nearly the height of the dam (see Figure 16-6). Intake vents near the lake's surface admit water that falls within large conduits, ultimately to strike turbine blades, which causes them to rotate at high speed and operate an electrical generator. HEP may be considered perennial because as long as the sun shines (practically eternity), water will be evaporated, fall to Earth as precipitation, collect in rivers, and be available for power production.

Figure 16-6: Dams may produce hydro-electric power while promoting flood control, irrigation, recreation, and drinking water supply.

HEP supplies about a quarter of the world's electricity. Not only is it a non-polluting energy source (nothing is burned), but the reservoirs may provide recreational opportunity, flood control, and water for people and agriculture. In addition, HEP projects may have important symbolic value. For example, China is presently engaged in construction of the giant Three Gorges Dam on the Yangtze River. While the finished project will provide all of the benefits noted previously, the dam is perhaps even more important (to the Chinese at least) as a symbol of the nation's emergence as a major world power and of the government's ability to accomplish great deeds.

On the negative side, construction of HEP projects is expensive. Many millions (even billions) of dollars may be required before the first bit of electricity is produced. Indeed, this is doubly bad because the areas of greatest hydroelectric power potential includes countries in Africa and Southeast Asia that do not have the financial wherewithal to fund such projects.

HEP development has additional geographical constraints. Obviously, you can't build a dam just anywhere. You need a river valley, and especially a narrow one for the simple reason that building across a narrow valley is cheaper than building across a wide one. Also, the eventual reservoir-related flooding upstream may cause serious displacement of people, agriculture, and transportation systems, all of which may occur at great cost.

Trading-off Resources: The Consequences of Resource Use

Further complicating resource geography is the fact that production and consumption of one natural resource may occur at the expense of another. The result may be deterioration of environmental conditions over specific geographic areas. Here are three examples from a wide range of possibilities:

- **Fossil fuels and air quality:** Americans desire low-cost access to gasoline, which they consume in massive quantities. Americans also desire clean air. The atmosphere, unfortunately, is the gathering place of pollutants that result from gasoline consumption. As a result, urban areas in particular, with their high concentrations of motor vehicles, often suffer high levels of atmospheric pollutants.

- **Fossil fuels and land subsidence:** Louisiana has the dubious distinction of being the only state that is measurably shrinking. The culprit is extraction of oil and natural gas in coastal parts of the state, which is causing the land to subside. As a result, coastal habitat and the land itself is being lost. Similar topographic events are occurring in other areas, though not necessarily with the same dramatic results.

✔ **Strip mining and soil loss:** In some areas large and valuable coal deposits are found just below Earth's surface. The safest and most economical way to extract these resources is to strip away the *overburden* (the soil and rock between the surface and resource) and then dig up the coal. This process is called *strip mining,* and its by-product is a deeply scarred surface (see Figure 16-7).

Formerly, strip-mining companies were allowed to do their thing and move on without devoting any effort to repair the damage. Now, however, laws require major land reclamation. But even the best repair job cannot fully restore some lands to their former status. In a rather cruel twist of geography, some of the country's most valuable coal reserves lie underneath some of the best Midwestern farmland. Fortunately, land reclamation has been fairly successful in that region, but full repair may take several years, and does not always result in land that is as productive as it was before. Some people question whether the trade-off of coal for soil is worth it.

Figure 16-7:
Strip mining disrupts and scars the landscape.

The global appetite for resources, and particularly energy resources, is very uneven. The United States alone accounts for about 26 percent of global energy consumption despite being home to only about 4 percent of the world's population. All of the countries of South America combined, in contrast, account for only 3.6 percent of global energy consumption, while those of Africa account for 2.6 percent (see Figure 16-8). Its economy and standard of living, as opposed to sheer number of people, explain the high consumption figure for the United States. Thus, it consumes 8 times as much energy annually than does India, which has more than 3 times as many people.

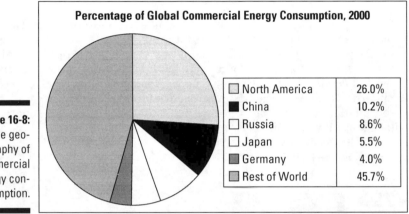

Figure 16-8:
The geography of commercial energy consumption.

Percentage of Global Commercial Energy Consumption, 2000

North America	26.0%
China	10.2%
Russia	8.6%
Japan	5.5%
Germany	4.0%
Rest of World	45.7%

Some people ponder the ethics of so few Americans consuming so much of the world's energy. But perhaps the more critical consideration is all of the countries and peoples around the world who are striving to raise their economies and living standards, and in so doing emulate American-style resource consumption. Assuming they meet with at least modest success, and adding the virtual certainty of global population increase, the signs point to a growing appetite for resources in the years ahead and growing strain on the natural environment to provide them.

Chapter 17

CBD to Suburbia: Urban Geography

*I*f the year was 1790, then this could be a newspaper headline: 195,000 Americans Now Live in Cities!

Sounds preposterous, doesn't it? But in fact, in 1790, the first U.S. Census revealed that only 5 percent of the nation's 3.9 million inhabitants lived in cities, which actually were tiny islands in a sea of rural settlement and wilderness. In every census since, the urban percentage has marched steadily upward (as seen in Figure 17-1). In 1920, it exceeded the rural percentage for the time, and has never looked back. Today around 75 percent of Americans may be categorized as urban. That amounts to about 222 million people out of a total population of some 285 million.

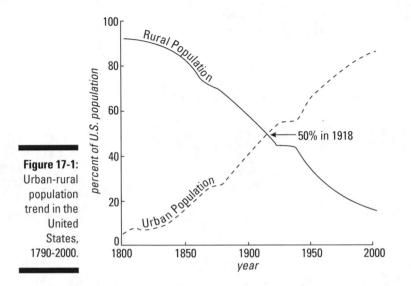

Figure 17-1:
Urban-rural
population
trend in the
United
States,
1790-2000.

Studying the Urban Scene

As centers of population, political and religious power, trade, and commerce, cities have always been geographically significant. At no time, however, have they been more important than they are today — not only because so many people live in them, but also because cities have grown in number and size. Literally millions of square miles of natural landscapes have become "cityscapes." Indeed, cities have become so expansive that they have complex geographies of their own. No wonder an entire sub-field called *urban geography,* the subject of this chapter, has developed to analyze:

- The reasons why cities develop at particular locations

- The factors that underlie urban growth

- The spatial arrangement of commercial, industrial, residential, and other kinds of land use within cities

- The planning that is meant to improve the quality of urban life

- The social and political tensions between people who live in different parts of cities, and between cities and fringing political jurisdictions

- The environmental quality issues occasioned by urban development and growth

Each of these will be touched upon to some extent in the pages that follow.

What does "urban" mean?

Obviously, urban means a lot of people bunched together in a recognizable town or city. But how many? The answer varies from one country to the next. As far as the folks who run the U.S. Census are concerned, the threshold figure is 2,000. Thus, if you have a town with that many people, then its residents are considered *urban*. If a neighboring town numbers 1,999 souls or less, then they are classified *rural*. Depending on where you live, "urban = 2,000 people" may make a lot of sense or sound like a bad joke. Throughout the lightly populated Dakotas, for example, the definition seems entirely appropriate. In New York City, however, where you may have that many people and more in a single building, the number 2,000 may seem woefully lacking.

Getting a Global Perspective

At the global scale, urbanization has broadly mimicked the American experience. No worldwide census data are available from back in 1790, but available anecdotal and statistical evidence suggests that the urban percentage for the world as a whole was about that of America's. Today, some 46 percent of the world's people live in urban areas, but that number masks significant variation. In the more developed world, 75 percent of the populace (exactly the American percentage) live in cities, although in some countries, the figure is much higher. Urbanization in Uruguay and Argentina, for example, is currently at about 92 percent and 90 percent respectively.

On the other hand, in the less-developed countries the rate of urbanization today is about 40 percent. It makes sense that this figure is so much lower because by definition a less-developed country is one whose economy is based on agriculture — so the populace tends to be down on the farm instead of downtown. But that statistic also masks variation. In Mongolia, for example, 57 percent of the people are urbanized, while in Bangladesh, the figure is 21 percent. Nevertheless, some of the world's largest cities have sprung up in developing countries, and with them some of the most challenging venues regarding urban planning and quality of life issues. Table 17-1 shows the top 20 largest areas. (Note: City is defined in Table 17-1 as a contiguous metropolitan area. That may include any number of municipalities that fringe a central urban area.)

Table 17-1	The 20 Largest Metropolitan Areas in 2000
City	*Population*
1. Tokyo, Japan	27,900,000
2. Bombay, India	18,100,000
3. Sao Paulo, Brazil	17,800,000
4. Shanghai, China	17,200,000
5. New York, U.S.	16,600,000
6. Mexico City, Mexico	16,400,000
7. Beijing, China	14,200,000
8. Jakarta, Indonesia	14,100,000
9. Lagos, Nigeria	13,500,000
10. Los Angeles, U.S.	13,100,000
11. Calcutta, India	12,700,000
12. Tianjin, China	12,400,000
13. Seoul, South Korea	12,300,000
14. Karachi, Pakistan	12,100,000
15. Delhi, India	11,700,000
16. Buenos Aires, Argentina	11,400,000
17. Manila, Philippines	10,800,000
18. Cairo, Egypt	10,700,000
19. Osaka, Japan	10,600,000
20. Rio de Janeiro, Brazil; Dhaka, Bangladesh	10,200,000

Source: The New York Times 2001 Almanac

Economic development brings with it employment prospects that are largely urban-based (for more information on this, see Chapter 15). Now, in developed countries these opportunities tend to be widespread and manifested by country-wide urbanization. In developing countries, however, they tend to be much more concentrated in the geographic sense with the result that the populations of one or two cities (the principal development centers) skyrocket. That is the reason why, as Table 17-1 shows, more than half of the world's twenty largest cities are in developing countries

Because of concentrated urban growth, an important aspect of Third World urbanization is the disproportionate occurrence of *primate cities*. No, this has nothing to do with *Planet of the Apes*. A city is classified as primate when it is far and away the largest city of a country. Classically, it tends to be at least twice as large as the second largest city. More often than not, it is also the capital. Thus, it stands as the overwhelming center of population, political power, employment, commerce and wealth.

Primate cities are powerful magnets for rural-to-urban migration, often resulting in sprawling, disease-ridden slums inhabited by poor folk who have come to seek work. Also, the concentration of so much power and prospective opportunity in a single setting may result in domestic political tension that pits inhabitants of a primate city against people in other parts of the country that feel deprived of their fair share of economic opportunity. On balance, therefore, the consequences of having a primate city are not entirely positive.

As suggested by the 1790 census data at the start of this chapter, the modern reality of so many people worldwide living in so many cities is at odds with the way things used to be. In fact, urbanization is at least a 5,500 year-old process whose early geography may be traced to a handful of locales.

Getting Started: Urban Hearths

Cities originated independently in five regions (shown in Figure 17-2) often called *urban hearth areas* — places where the first cities developed. They include:

- Mesopotamia: The Tigris-Euphrates River valley that mainly lies in modern-day Iraq.

- Nile Valley: Especially in what is now Egypt.

- Indus River Valley: In what is now Pakistan.

- The Lower Huang Ho (Yellow River) Region: This is located in Northeastern China.

- Mesoamerica: The region that extends northward from Nicaragua to and including Central Mexico.

Mesopotamia probably has the honor of being the first. Its oldest urban ruins date back about 5,500 years. Ancient Egypt appears to be a close second.

Trade and commerce played important roles in the growth and development of hearth-area cities. But so, too, did the need for centers for religious rituals and observances. As a result, imposing structures of ceremonial significance (such as pyramids and temples) dominated "skylines." These tended to be located alongside or at the intersection of broad boulevards aligned in directions that often had theological significance. Complementing these imposing structures were achievements of more mundane, yet fundamental, varieties. Yesterday's cities (yes, even without traffic lights and fast-food restaurants, they're still considered cities), with populations that probably didn't exceed 25,000, showcased distinct residential, commercial, and institutional zones, as well as municipal services such as water supply and garbage removal. These prove that municipal planning and governments were at work even back then. Thus, the urban hearth areas were scenes of important experiments and innovations in human settlement, the results of which were passed down to succeeding generations, modified, and ultimately spread to other lands.

Geographically, what is perhaps most important about these hearths is their locations. Each coincided with an area of high agricultural productivity. This positioning made a great deal of sense because the principal prerequisite for the appearance and growth of cities is an agricultural surplus. Why? Because that alone frees some people of the necessity to produce food and allows them to become artisans, traders, shopkeepers, soldiers, clerics, and so on, and cluster in settlements — cities — that promote the interaction of those services. Accordingly, each hearth (except Meso-America) is located amidst fruitful lands that roughly parallel one or more large rivers. The richness of the soils, whose fertility was regenerated by silts deposited by periodic floods, allowed people with even the most basic agricultural technology to produce good harvests (food surpluses) on a regular basis. Each hearth (including Meso-America) was also noted for impressive irrigation systems that maximized the agricultural potential of local waters and soils. Given the significance of these water systems, whether natural or human in origin, the early cities and their peoples are sometimes referred to as *hydraulic civilizations*.

Finding Sites for Cities

While the civilizations that produced those early cities are things of the past, the locations of urban hearth areas demonstrate a geographic principle that has persisted through time. Namely, cities don't just happen. Instead they tend to spring up in particular locations because of some characteristic(s) of place that was attractive to settlers.

For example, and strange as it may seem today, the Spaniards settled the Los Angeles, California area for its agricultural potential. Mountains and foothills come down to the sea along the West Coast, affording limited venues for agriculture and settlement. Los Angeles area, however, offered a substantial expanse of relatively flat land complemented by a then-adequate local water supply that trickled down from the fringing San Gabriel Mountains. That was then. Today, few, if any, people live in LA because of its crops potential (legal ones, that is), and that makes an important point about urban location factors. Namely, the reason why a city began at a particular location may have nothing to do with the reasons why people live there today.

The list of factors that explain why cities began and developed at particular sites is lengthy. The four that are discussed in the following sections are the most popular.

Confluence

A *confluence* is a place where two rivers meet to form a combined flow. Pittsburgh, Pennsylvania, for example, began as a small settlement where the Allegheny and Monongahela Rivers join to form the Ohio River. Other examples are St. Louis, Missouri, near the point where the Missouri and Mississippi join waters; Khartoum, Sudan, where the White Nile and Blue Nile come together; and Manaus, Brazil, where the Amazon and Rio Negro combine their waters.

Rivers served as highways in olden times, so being at a confluence was like being at the hub of a transportation network. Confluences are prone to flooding, however, so availability of high ground is essential for safe settlement. Fortunately, plenty of high ground is available in Pittsburgh. In contrast, the land is rather low-lying where the Mississippi and Missouri meet. For that reason, St. Louis is not located right at that confluence, but a few miles downriver, where the banks are appreciably higher.

Protected harbor

Ever try to load or unload a sailing ship that is pitching to and fro in moderate-to-heavy waves? I didn't think so, and don't bother trying. It's very difficult and downright dangerous. What you need are calm waters, preferably in a *protected harbor* setting, where intervening land negates the big waves of the open sea. Because shipping has long been a principal means of moving people and goods, protected harbor settings have long been prime sites for settlement. Geographically, what do Boston, New York, Baltimore, Norfolk, San Diego, San Francisco, Seattle, Vancouver, Tokyo, Sydney, Mumbai (Bombay), Karachi, Hamburg, Havana, and Rio de Janeiro have in common? You got it.

Head of navigation

A *head of navigation* is the farthest point upriver that a boat or ship can travel from a river's mouth. Washington, DC, for example, is located at the head of navigation of the Potomac River. That is, ships of good size can proceed up the Potomac from its mouth as far as the site of that city, where rapids make further navigation impossible. In other instances, waterfalls, shallow water, or narrowness calls a halt to water-borne traffic. In the settlement of North America, heads of navigation sometimes proved a logical spot for boats to stop, be unloaded, and later for settlements to develop. In some cases, these sites afforded other advantages for settlement. Where waterfalls or rapids impeded navigation, these features sometimes provided sources of waterpower for mills and other economic activities. Narrows or shallow water, in contrast, sometimes offered opportunity for bridge construction. Other examples of heads of navigation in the United States include Trenton, NJ (Delaware River); Louisville, KY (Ohio River); and St. Paul, MN (Mississippi River).

Defensive sites

Defensive considerations were paramount in the siting of several cities. The oldest part of Paris, for example, is the Ile de la Cite, a small island in the middle of the Seine River. The river served as a protective moat for the island's oldest known inhabitants, a tribe called the Parisii. Similarly, the oldest part of Montreal is an island in the St. Lawrence River that served a defensive purpose.

Some rivers exhibit dramatic U-shaped bends called *meander loops*. By siting a town on the inside of a loop, the river served as a protective perimeter. Examples of this include New Orleans, Louisiana, and Berne, Switzerland.

Acropolis sites, characterized by locally high ground, which is fairly easy to defend, have also been favored. Examples include Quebec City, Albany (New York), and Athens, Greece — whose famous Acropolis is the source of this category. *Peninsula sites* were also favored because a wall across a neck of land, complemented by the surrounding water (a natural moat), greatly facilitated defense. Boston and Bombay, also noted for their natural harbors, are good examples.

Because of these and other site-related factors, settlements developed at various locations around the world. All were small at first. Many did not stay that way.

Getting Big: Urban Growth

At favored sites around the world, hundreds (if not thousands) of once-humble settlements have developed into cities of varying size. And the process continues. In fact, urbanization is not just a major human characteristic, but also a major trend. But the pace of urban growth and the reasons for it are not the same everywhere. Because space does not permit a truly global treatment of this truly global process, I'm going to focus on a prime example of the urbanization process — the United States.

Much of the growth of American cities has been due to immigration and natural population growth. But other factors have also encouraged cities to expand not just population-wise, but also with respect to the physical area taken up by them. The following sections discuss some of the principal factors.

Rural-to-urban migration

Over the world as a whole, the last century has witnessed substantial rural-to-urban migration. In the United States and several other countries, a couple of processes have been fundamental to this shift.

Mechanization of agriculture and farm consolidation

Agriculture was once a largely manual enterprise that required many hands to tend a farm of decent size. But with the advent of farm machinery (tractors in particular), the number of people needed to do farm work was greatly reduced. At the same time, mechanization also encouraged farm consolidation (in which one farmer buys out another, thereby increasing the acreage owned by the buyer) because tractors made it possible for a farmer to farm much more land than was the case before. Thus, the pre-mechanization rural landscape that consisted of a large number of small farms was transformed into one containing a comparatively small number of large farms. In 1920, (largely pre-tractor times) 6.5 million farms in the U.S. averaged 149 acres. Now about 2.1 million farms average about 430 acres. During the same period, farm population dropped from 31 million to 4 million. Cities absorbed the brunt of the difference.

Changing economy

The mechanization of agriculture was symptomatic of a structural change in the overall economy that witnessed an increase in industrial and commercial enterprises that were largely urban-based. Manufacturing was particularly important because it afforded job opportunities for unskilled or semi-skilled individuals who were either recent immigrants or recently forced off the farm. Thus, in 1870 the overwhelming majority of African-Americans were rural Southerners. Less than a century later, a majority were urban dwellers, many of them holding factory jobs in cities outside of the former Confederacy.

Changing means of transportation

How big would a city be if every inhabitant had to walk to work? The answer is pretty simple: Not very big at all. And in fact, that's the way things were long ago. When overland transport was exclusively a matter of walking, horseback, carriage, and cart, it took a while to move people and products meaningful distances. As a result, cities tended to be rather compact.

Over the years, however, changes in transportation have allowed people to live progressively farther from their places of employment, thereby expanding the physical size of cities. Following are various "urban transportation ages," though not all apply to every city.

Pre-mechanization

Prior to about 1890, foot, carriage, or horse-drawn trolley accomplished most movement of people and goods. This resulted in compact cities, the largest of which could be completely traversed in a 30 to 45 minute walk. Attached or closely-spaced housing was the norm.

Electric trolleys

Around 1890, introduction of electric-powered trolley cars provided a quantum leap forward in speed. While a 30-minute walk might allow a commuter to live perhaps as far as 2 miles from work, a commensurate trolley ride could extend the range of residential potential to 5 miles and more. As a result, the outer edge of urban areas expanded significantly.

Commuter railways

Starting around 1900, commuter railways developed using standard, instead of narrow, gauges. The result was another quantum leap in speed that expanded the range of commuting to 20 miles and more. Commuter trains can only travel where tracks are present, of course. Thus, cities expanded along rail corridors that were rather like tentacles, leaving countryside in between.

Pre-freeway

Advent of mass automobile ownership facilitated unprecedented freedom of choice with respect to residential location. Suburbs consequently developed and "filled in" substantial portions of rural land between the commuter rail tentacles described previously. Until roughly the mid-1950s, however, highways were largely of the 2- and 4-lane variety. Combined with red lights, stop signs, and traffic, commuting speed was somewhat limited and so too, therefore, the distance at which one may choose to live from the city.

Freeway

Multi-lane, high-speed, limited access (as in limited points of entrance and exit), divided highways first appeared in the early 1950s and, in the U.S., proliferated with the advent of the federally funded Interstate Highway System. Initially, at least, the effect of these arteries was to greatly increase the average speed at which a motorist could travel and therefore also increase the potential distance from residence to job. Complemented by construction of *beltways* — interstate-style ring roads designed to help people travel around the central city without going through it — this ongoing age has witnessed significant increase in the "filling in" of remaining rural land between commuter rail "tentacles," and pushed the outer edge of the city in multiple directions, as shown in Figure 17-3.

Automobile ownership

By themselves, neither freeways nor roads of other sorts tell the full story of contemporary urban growth in the U.S. More telling is the number of people who own cars and use those roads and freeways to commute to work from a suburban residence. And that number is humongous. More than 130 million automobiles are registered in the United States and another 80 million trucks of various sorts. According to the 2000 U.S. Census, more than 90 percent of households own a motor vehicle, and 18 percent of households own three or more.

Figure 17-3:
Freeways have been a key element in urban expansion in the U.S. and some other developed countries.

Low-cost fuel

The price of automobile fuel in the United States is relatively low. That prompts urban sprawl by encouraging automobile usage and increasing the distance people are willing to commute. Now, some people may recoil at the notion that gas is cheap in the United States. All they need to do, however, is vacation in Europe or almost any other part of the world to find out firsthand just how inexpensive American gas is. The principal reason for the price difference is low taxes on gas consumption in the U.S., which encourage automobile purchases, new home construction, and other consumer activity that contributes to the economy even as it contributes to urban sprawl.

Home mortgage deductibility

The U.S. federal government has long encouraged home ownership by allowing buyers to deduct from their federal income tax the interest paid on home mortgages. The effect is to encourage new home construction and urban sprawl. Canadians, in contrast, do not get the same tax benefit. Their cities also sprawl, but generally not to the extent of those on the U.S. side of the border. Several factors explain that, and mortgage deductibility is one of them.

Looking Inside the City

In addition to sites and sizes, cities around the world vary dramatically in their internal physical appearance. Numerous cities in Europe, North Africa, and Asia, for example, contain an "old town," characterized by traditional architectural styles, and a citadel, cathedral, or mosque that towers above historic low-rise buildings. In other cities, strict building codes produce an air of distinctness. Thus, while many modern cities are defined by center-city skyscrapers, Paris and Venice, to take just two of many examples, have uniformly low skylines that preserve historic character and visual dominance of leading landmarks.

Internal physical appearances of cities may also vary as a function of wealth or income of residents. Urban neighborhoods of the rich and poor not only look different, but also tend to occupy different locations in the cities of different countries. Thus, in many developed countries, the urban poor tend to live in older housing in the inner city. The wealthy, in contrast, generally tend to live in the suburbs or urban fringe. In many developing countries, however, the pattern tends to be the opposite. That is, the poor tend to concentrate in slums that are located on the urban periphery while the wealthy are located much closer to the urban core. Accordingly, go up into the hills on the outskirts of many American cities and you are likely to see expensive homes that afford a great view. Do the same in most large Latin American cities, and you are likely to see some of the worst slums and shanties imaginable.

Distinct areas of commercial, industrial, residential, and other kinds of land use came into being as cities grew. Their extent and arrangement vary substantially in different cities in different countries. In the United States, the factors promoting urban growth are largely the same throughout the country. To a fairly substantial extent, therefore, American cities exhibit a certain sameness with respect to their structure.

Specifically, at the center there tends to be a discernible "downtown" dominated by relatively tall building and mostly non-residential land use. On its fringe are high-density residential neighborhoods characterized by high-rise apartments and attached multi-family dwellings. As distance from downtown increases, population density generally decreases. Detached housing becomes the norm and lot sizes increase. Mixed here and there amidst suburbia are shopping centers of varying sizes, with occasional full-blown shopping malls on the periphery of larger cities.

If you live in a city, or are familiar with one, how well does that description conform to your experience? The sections that follow consider urban parts and processes in greater detail.

The Central Business District (CBD)

At or near the center of most cities is what students of urban geography call the *Central Business District* (*CBD*) and what normal, well-adjusted people call "downtown." This is the core of the city and in most instances is characterized by:

- ✔ The area where the city began
- ✔ The focal point of the main transportation arteries
- ✔ The tallest buildings
- ✔ City hall, court buildings, and government offices
- ✔ Major commercial, retail and office buildings

In some respects "downtown" is preferable to "CBD" because the former infers that much more is going on in the area than just business. Still, the concentration of business enterprises is such that commercial land use dominates the CBD and gives rise (quite literally) to its most outstanding characteristic: large and imposing structures, usually including the tallest buildings in town. Competitive bidding for real estate is what leads to those structures, and both the bidding and the buildings deserve at least modest description.

Competitive bidding and the rent gradient

Historically, downtown was *the* place to be as far as commercial enterprises were concerned. That is because it was the focal point of the transportation network and therefore readily accessible to a high percentage of the city's population, all of whom were potential customers or employees. As a result, demand for downtown business locations quickly exceeded the supply of real estate. When demand exceeds supply, whether you're talking real estate or rock concert tickets, the price rises. This *competitive bidding* for downtown real estate drove up land prices in that area relative to other parts of the city. As a general rule, competitive bidding for land generally decreases as distance from the urban center increases. As a result, land values tend to "peak" downtown and then decrease as distance from the CBD increases — sharply at first, and then more gently (as illustrated by Figure 15-10). Geographically, therefore, land values exhibit a gradient that declines from the CBD to the periphery of the city. The trend line is often called the *rent gradient*, even though land purchases rather than renting *per se* is the predominant financial transaction.

Tall buildings

The quintessential result of the downtown peak in the rent gradient is tall buildings. And for good reason. Suppose you have just purchased land downtown and borrowed a hefty amount of money from a bank to do it. Now you've got a loan to pay off, along with its accruing interest. Also, given the high price you paid, you can bet your last dollar (if you have one) that the annual real estate tax on your property will be through the roof (if you can afford one).

How can you cope with this predicament and make a tidy profit to boot? Create as much floor space as possible on your property and rent it out to businesses that will pay a pretty penny to locate downtown. And the way to create that floor space is to go vertical — that is, build as many stories (thereby creating as much rental space) as you and your bankers deem feasible. In most cities, you find some tall buildings that are not downtown, but the rent gradient usually guarantees that the CBD will have the highest concentration of them, an example of which is shown in Figure 17-4.

Figure 17-4: Tall buildings express the rent gradient that peaks in downtown Chicago.

Residential areas

Urban residents generally live in one of two areas: multiple housing buildings that fringe the CBD or detached, single-family residences of outer-city areas. The following sections discuss some of the areas that crop up when people choose to live in and around cities.

Rich folks-poor folks, suburbs-inner city

As a general rule (which has numerous exceptions) people of low-to-modest incomes tend to live in inner-city high-density neighborhoods while the more affluent reside in the suburbs. This geographic pattern is not as old as the cities themselves, but instead evolved over time. Specifically, an old high-density residential area usually adjoins a city's CBD. Most often the buildings date from pre-automobile times, when the need to be close to work encouraged dense settlement. As affluence rose and transportation improved, the middle- and upper-income people (largely white) tended to move to outlying areas, while low-income people (largely minority) moved into the newly vacant residential spaces. As the number of inner-city poor people rose, landlord incentive to spend money to maintain and improve their housing declined. Slums were the virtually inevitable result despite occupying land little more than a stone's throw from high-value CBD real estate.

Other factors have subsequently served to place suburban real estate beyond the reach of the inner-city less-well-off. These include local (suburban) decision-making not to build public or low-cost housing, and passage of ordinances that mandate large minimum lot sizes or house sizes. Also, lack of efficient (or any) public transportation in suburban areas decreases the likelihood that the poor will move in.

As a result, cities and suburbs generally end up being inhabited by people of different race and social class. These people tend to have different needs, different attitudes toward social services, different views on the role of government, different perspectives on taxation, and different priorities about how tax money should be spent. Political tension between inner city and suburban residents is an almost inevitable side effect. This sometimes results in one political party being "the party of the city" and another being "the party of the suburb."

Ethnic neighborhoods

An *ethnic neighborhood* is a residential area in which people of common origin voluntarily live in close proximity to each other. (This is in contrast to a *ghetto*, in which the choice of residential location is involuntary.) These are common features of American cities (Canadian and European ones, too) inasmuch as they afford foreign immigrants opportunity to

- ✔ Interact with others who literally and figuratively speak their language
- ✔ Maintain contact with kinfolk
- ✔ Have access to stores and eateries that specialize in familiar goods and foodstuffs
- ✔ Worship in a preferred manner

The vast majority of these neighborhoods are found in high-density inner-city residential areas. That is a reflection of the (normally) limited disposable income of immigrants, proximity to real or potential places of work, and, again, strong desire to be among people with a similar background.

Applied Geography: Lessons from Bhopal?

On December 2, 1984 an explosion at a chemical plant in Bhopal, India produced a cloud of toxic hydrogen cyanide gas that settled over nearby residential areas. Some 5,100 people died, and many thousands more suffered permanent injures to lungs and eyes. Subsequent investigation suggested the tragedy could have been prevented by modest investments to improve operational safety within the plant. On the other hand, and not to make light of those findings or of the company's culpability, some have suggested the extent of the disaster was due to a lack of zoning.

Zoning is the process in which urban planners (many of whom are trained in geography) allocate particular kinds of land use to particular parts of a city. The desired purpose is to promote smooth functioning and livability. With respect to Bhopal, why, some have asked, was a factory of that sort and a densely populated residential area ever allowed to exist side-by-side? An appropriate degree of geographic separation between plant and people, consistent with rational zoning, would not have prevented the accident, but no doubt would have resulted in far fewer casualties.

Suppose a manufacturer of toxic chemicals wanted to build a plant in the city or town where you live. Also assume that you are the local urban planner (Congratulations!) and that the plant's coming to town is a done deal because its economic benefits are too good to pass up. The only remaining question is: Where should it be located? What's your opinion? And how would you respond to critics of zoning who contend that property owners should be allowed to do whatever they please with their real estate (including building a chemical plant on it) without first securing the approval of people like you?

Ghettoes

As previously mentioned, a ghetto is an involuntary neighborhood where members of a group are forced to live. Typically, they are found in an inner-city, high-density residential area. *Ghetto* is Italian for "foundry," and reflects the fact that Jews in Venice of old were relegated to (and literally locked in at night) a small area in the factory district.

Opinion varies (often heatedly so) about the appropriateness of the term in America, where it has been applied most often to the African-American and Asian-American urban experiences. While these peoples have never been literally locked in a particular area, in certain times and cities, they have been victims of *red lining*. This is a discriminatory and now illegal practice whereby real estate agents mark off on a map (classically with a red pen) parts of town in which housing may and may not be sold to members of these groups. The result, for all intents and purposes, is a ghetto.

Leaving Downtown, Living Downtown

In recent decades two rather opposing processes have impacted the character of cities. One involves movement to the suburbs and outer fringe of large stores and other entities formerly located downtown. The other (and more recent change) entails the movement of middle- and upper-income people into downtown areas. Each process, described below, has brought significant socioeconomic change — for better or for worse — to affected areas.

Moving out of downtown

Decentralization refers to the appearance in the outer city of functions and land use formerly associated almost exclusively with the CBD. The quintessential example is a large downtown department store that closes, perhaps after years of being part of the CBD, and relocates to a newly opened suburban shopping mall. The most profound effect is to move commerce and retailing from the center city to the outer city (hence decentralization), often resulting in vacant real estate downtown and loss of jobs for inner city residents. Here are some principal manifestations and results.

Suburban shopping malls

Enclosed shopping malls consisting of two or more major department stores and dozens of lesser-sized shops of considerable variety are perhaps the highest expression of decentralization (illustrated by Figure 17-5). They come about for one or more of the following reasons:

✔ The cost of land or rent is cheaper in the suburbs than downtown.

✔ Proximity to people with high disposable income stimulates business, and more people fit that description in the suburbs than in the central city.

✔ Malls are more convenient to get to by auto than is downtown. Specifically, getting to a suburban mall is likely to involve less traffic, fewer lights, and lots of free parking.

✔ Shoppers (especially suburbanites) tend to prefer mall shopping because of the juxtaposition of different kinds of shops, their comparative cleanliness and perceived safety, and their enclosed climatically controlled environment.

Office parks

An *office park* is a cluster of multi-story office buildings (glass facades are particularly popular) that typically occupy more than enough acreage at or near a major suburban intersection. Trees and other landscaping (hence, park) complement the architecture. The suburban location often constitutes a preferable commute for employees who might otherwise have to drive downtown. Similarly — and just like shopping malls — the locations facilitate accessibility by existing near potential middle- and upper-income clients, who are more likely to reside in the outer city than the inner city, and who prefer the former setting to the latter as a place to take their business.

Edge cities

Comparatively small *edge cities* sometimes appear on the outer fringe (or edge) of large cities. Sometimes this is simply the result of merging together due to urban growth. More often, however, edge cities are newly incorporated areas that arise in response to the needs of suburbanites for a wide range of goods and services, and their preference to obtain them in car-friendly places that are closer to home than the old CBD.

Jurisdictional spillover

Urban sprawl often results in a contiguous urban area that exceeds the original city limits and spills over into the territory of other cities, counties, or even states. As a result, hundreds or even thousands of people may earn a paycheck in one jurisdiction, but pay taxes and spend their disposable income (thereby encouraging job-creation) in another jurisdiction. This is cool as long as it all "evens out," but it seldom does. The jurisdiction that "loses" is the one with the highest number of workers who live somewhere else, and typically that is the central city.

Figure 17-5:
Suburban shopping malls like this one typify decentralization.

Moving back downtown

In recent years, a rather provocative geographic trend has been occurring that rather flies in the face of urban sprawl and decentralization. Called *gentrification,* it's the process by which middle- and upper-income people (who you would normally expect to see settling down in the 'burbs) buy and refurbish central-city housing.

Causes

The causes of this are diverse, but typically include one or more of the following:

- ✔ Preference for social diversity and cultural amenities, which typically are in greater supply in the CBD as opposed to suburban settings.

- ✔ Desire by people whose jobs are downtown to forgo the time and cost of commuting from the urban fringe.

- ✔ The often favorable cost differential between buying and refurbishing inner city property versus the purchase of property with a home on it in the suburbs.

- ✔ Availability of large rental spaces downtown pursuant to commercial and manufacturing concerns that have decentralized or gone out of business.

Characteristics

Having been set in motion, gentrification may be manifested in ways that may include the following:

- ✔ Social change takes place in the inner city gentrifying area as more affluent people move in.

- ✔ Housing stock becomes refurbished.

- ✔ The continued influx of buying power leads to competitive bidding, which drives up housing and residential rent costs.

- ✔ Spending by newcomers has a spillover effect, increasing the value of commercial real estate as landlords find they can command higher rents.

Consequences

Once it has firmly taken hold, gentrification may have significant consequences that include the following:

- ✔ Rising rents cause displacement of long-term commercial and residential tenants, giving rise to homelessness.

- ✔ Low-rise buildings are bought and demolished to make way for high-rises.

✔ "Mom and pop" stores become boutiques. The former unique commercial character of the area becomes increasingly "placeless" as chain stores and franchises become more a part of the local landscape.

✔ The urban tax base expands.

The bottom line is that gentrification is a double-edged sword. It encourages homelessness and social displacement even as it encourages revitalization of the inner city and reinvention of the old CBD.

Facing up to Environmental Issues

Just as city growth has had diverse social and economic consequences, so, too, has it given rise to a host of environmental challenges. The following sections cover a few of the biggies.

The urban heat island

The atmosphere over a city tends to be significantly warmer than over rural areas, a phenomenon called the *urban heat island*. While the effect may be a modest blessing during the cold of winter, it can also amplify summer heat waves, resulting in higher rates of health emergencies and heat-related fatalities than in non-urban areas. Here are three reasons for this phenomenon:

✔ Asphalt, concrete and other hard materials that dominate city surfaces have rather high capacities to absorb solar energy, causing urban areas to warm up much faster than countryside.

✔ Heat is a by-product of motors and machines, which abound in cities. Ironically, air conditioners are a major contributor to urban heat islands during summer.

✔ Exhaust from automobiles and engines is warm in and of itself, but also has high concentrations of gases that absorb and retain solar energy with great efficiency.

Dust domes and smog

Exhaust from internal combustion engines and other sources is high in particulate matter and smog-producing chemicals. As a result, air pollution problems, attendant to resulting *dust domes* and *smog,* tend to be particularly acute in urban areas, as manifested in part by disproportionately high rates of asthma and other respiratory problems. The flip side of sorts is that dust particles serve as the focus for condensation, leading to production of rain. Thus, urban areas tend to have higher precipitation rates than neighboring rural areas, the rainfall "flushing" the heat and smog from the atmosphere and providing temporary relief from the heat island effect.

The extent of dust domes and smog vary from city to city, and not simply as a function of sizes. Terrain may be a factor. Mexico City and Los Angeles, for example, occupy low-lying lands that are surrounded and partially surrounded respectively by high hills and mountains that contain and concentrate air pollution. Existence and enforcement of environmental law, particularly as it relates to engine emissions, is also a factor.

Finding a place for refuse

Large cities produce large quantities of refuse and sewage and have few solutions regarding their safe disposal. Pursuant to passage of environmental protection laws, ocean dumping is no longer an option, and many cities are running out of sites for landfills within their boundaries. One of the more provocative consequences has been substantial interstate commerce in refuse, in which typically a poor rural county agrees (for a price) to serve as the final resting place for big-city garbage and sewage. Increased use of efficient, high-tech incinerators within cities themselves may be the way of the future, provided they comply with environmental protection laws designed to promote clean air. Even if that happens, however, the contribution of this technology to dust domes, smog, and urban heat islands will be measurable.

Chapter 18

Impacts on the Environment

· ·

· ·

*T*he Chesapeake Bay is the largest *estuary* (an inlet where the salty tide meets freshwater current) in the United States and perhaps the most economically important one, too. The greatest source of oysters in the U.S., Chesapeake Bay is also the world leader in blue crab production. Striped bass, perch, shad, flounder, and a host of other species help comprise this historically rich and robust fishery. So, naturally, it's rather disturbing when fish by the thousands suddenly turn up dead.

The immediate culprits are algae, tiny aquatic plants that takes in oxygen from the water. Occasionally, algae reproduces by the gazillions, creating algal "blooms" that deplete oxygen over large portions of the Bay and suffocate large numbers of fish. These blooms are caused by periodic build-up of phosphate and nitrate in the water. Algae, it seems, just love these substances, and reproduce like crazy in their presence.

That, of course, leads one to inquire into the origins of the bloom-producing chemicals. The Bay is not an isolated body of water. Instead, rain and runoff connect it with a broad land expanse (see Figure 18-1). The phosphates originate mainly in sewage treatment plants that empty into rivers and streams that empty in turn into the Bay. Nitrates originate mainly in agricultural fertilizers that run off from farms into the same Chesapeake tributaries. Indeed, virtually hundreds of watercourses from six states drain into the Chesapeake, collecting phosphates and nitrates from many, many sources along the way.

Maybe you're saying, "OK, but what does this have to do with geography? This sounds to me more like chemistry or biology." Geography is concerned with human and natural phenomena that give character to different parts of Earth's surface. Often, of course, important characteristics that we observe in a location or region are products of interaction between people and nature. While these relationships are sometimes harmless, they may also have negative consequences, as in the Chesapeake Bay example.

Environmental geography is the sub-field that focuses on human impact on the natural environment, and is the topic of this chapter. Previous chapters have referred to negative environmental impacts without mentioning environmental geography by name. Devoting an entire chapter to it is meant to emphasize both the magnitude of society's ability to alter the face of the Earth, and the existence of a discreet area of geography that is devoted to the topic. Although ending the meaty part of this book with talk of pollution and such may seem like a downer, it affords the opportunity to think critically about the role of geography in assessing the present and planning for the future.

Figure 18-1: Hundreds of watercourses from six states drain into the Chesapeake Bay.

Grasping the Basics — Environmentally Speaking

Environment refers to the myriad of natural characteristics and conditions that affect and are affected by humans. *Environmental geography*, therefore, is the study of the characteristics of locations and regions that are the result of human-nature interaction. This is closely allied to the field of *ecology*, which studies the complex connections that link the following elements:

- ✔ **Atmosphere:** the climate surrounding Earth

- ✔ **Hydrosphere:** water on Earth in all its forms

- ✔ **Lithosphere:** the solid earth including soil and other loose surface particles

- ✔ **Biosphere:** all life on Earth

- ✔ **Culturesphere:** meaning people, and emphasizing that it is by means of culture that different human groups interact with nature

The emphasis on location and place is what most distinguishes environmental geography from ecology. Thus, while an ecologist might study agriculture-nitrate-algae-fish linkages in the abstract or at the scale of microbiology, environmental geography studies them as large-scale events that characterize locations or regions. In that regard, geography is very much at home with the study of *ecosystems,* the living things that occupy particular areas together with the inorganic elements on which they depend. In Chapter 2, for example, there was a discussion of the African lion that basically was about an ecosystem. Specifically, lions (biosphere) inhabit areas where the climate (atmosphere) results in rainfall (hydrosphere) characteristics that produce grassland (biosphere). These grasses attract grazing animals (biosphere) like wildebeest, zebra, and impala that in turn attract lions. But the grasslands also result in soils (lithosphere) that are amenable to agriculture. As a result, people (culturesphere) have converted grassland to farms and grazing lands, depriving lions of the habitat they need to thrive.

Contributing Factors: Pollution on the Move

Impacts on the environment take several forms. In the sections that follow, several kinds are described together with natural phenomena that often (and perhaps surprisingly) help to turn small environmental problems into major ones.

Making an impact

Perhaps the most obvious human impact involves pollution, the introduction of substances (pollutants) that are harmful to the environment. These may result in *degradation,* the reduction in quality of natural environmental elements, or in *depletion,* the reduction in quantity. Human impacts can also be manifested by acts of removal or addition. Thus, the wholesale removal of forests (*clear-cutting*) may not only deplete forest resources, but perhaps also contribute to soil erosion. On the other hand, introduction into the environment of some element for which no natural counterbalance exists may have an effect no less profound than, say, accumulation of phosphates and nitrates in the Chesapeake Bay. A good example is provided by the kudzu, a fast-growing vine that was let loose in the South and has proceeded to dominate the natural vegetation over large areas.

Geographically, pollution may originate as a location-specific *point source* or as a larger scale *non-point source* (sometimes called an *area source*). In the Chesapeake Bay example, a pipe at a sewage treatment plant that discharges phosphate would exemplify point source. Crop fields in, say, Maryland, on which fertilizers that are high in nitrates have been applied, would typify non-point source or area source.

Spreading the mess

By themselves, pollution sources *per se* generate limited environmental impact. The major problem, rather, is that pollutants tend not to stay put because they are released within an environment characterized by several kinds of motion. In other words, nature has mechanisms that "spread the mess." As a result, pollution that starts out as a confined geographic event may end up affecting broad areas and having extensive consequences. Previous chapters have described these mechanisms in some detail. Here is a brief recap with examples.

The water cycle

The water cycle (see Chapter 8) is a major "culprit" for spreading the mess. Rainfall originates, falls to earth, and runs off the land to create streams that flow to the sea. In so doing, the water cycle acts as a pervasive and efficient mechanism that "spreads the mess."

This process picks up the phosphate discharge, carries it away, and turns a local event (discharge) into a problem of Chesapeake-sized proportions. This process also produces run-off on an agricultural field that picks up nitrates from fertilizers and carries them away. And of course, the Chesapeake Bay does not have *a* tributary, but instead more than a hundred of them, which together collect pollutants from thousands of point and non-point sources.

Wind

In innumerable instances, smoke and pollutants from a manufacturing or electrical generating plant go up a chimney (point source) and end up in the atmosphere. Here, too, the resulting environmental impact would be nominal if nature just stayed put. But, of course, it does not. Wind is another component of the dynamic environment. In Chapter 9, I discussed how solar energy leads to creation of high- and low-pressure systems that cause wind. And as a result, pollution that starts out as a very localized phenomenon becomes geographically general.

For example, as sanitary landfills fill up, incineration is increasingly looked to as a means of solving garbage disposal problems, especially in big cities. But even the most efficient incinerators generate pollution that is released to the atmosphere, and then spread over wide areas, thanks to wind. Thus, a source of point pollution (the incinerator chimney) may affect the health of humans, vegetation, and property values over a wide area.

Currents

In many times and places, oceans have been convenient receptacles for human refuse. Also, as the global economy has grown and become more inter-connected, maritime traffic has increased, and with it the possibility of incidents that result in the release of cargo that is harmful to the environment. At this point the familiar story line repeats: If nature stayed put, then the impact would be minimal. But we live in a dynamic environment. The oceans are restless. Surface currents (see Chapter 9) can carry things far and wide while tides and waves can affect every coastal nook and cranny. Once again, therefore, nature can spread the mess, so what starts out as a local event is turned into an issue of greater geographical proportions.

The possibilities were amply demonstrated on March 24, 1989, when the Exxon Valdez oil tanker went aground in Prince William Sound, Alaska, and ultimately spilled about 11 million gallons of crude oil into the sea (Figure 18-2). Although the Sound is a somewhat confined water body, the Alaska Current slowly and inexorably began spreading the mess — in this case, the oil slick — in a westerly direction. Coastline 400 miles away was eventually affected. However, given the numerous islands and inlets that characterize the region, some 3,000 miles of shore were contaminated. (Dead wildlife included more than 575,000 birds.)

Figure 18-2: The Alaska Current spread the Exxon Valdez oil spill over a great distance.

Bear bile, anyone?

One of the more intriguing human impacts on the environment concerns demand for animal body parts to serve as folk remedies. In China, for example, powdered tiger bone is the principal ingredient in a medicinal cocktail that is believed to relieve ulcers, malaria, rheumatism, and other ailments. A similar brew that uses ground rhinoceros horn is thought to heighten the sex drive and cure impotency. People use bile from the gallbladders of bears to treat liver disease, fever, hemorrhoids, and other ailments. Literally hundreds of millions (if not billions) of people belong to culture groups that swear by these age-old remedies.

Local demand for body parts has taken its toll on the local supply of desired animals. As a result, tigers and bears are nearly extinct in the wild in China. As numbers dwindle, prices skyrocket. Bear bile reportedly sells in China for as much as 15 times the price of gold. Tiger bone in parts of Asia sells for upwards of $500 per kilogram. As a result, captive breeding programs in China raise tigers for slaughter for bones, and bears for periodic "milking" of gallbladders. But even this cannot satisfy demand.

Body parts must be imported from somewhere else. And at prices like these, it's a poacher's dream come true.

Here is where geography fully enters the picture. A geographically local (Chinese) demand for body parts has generated a geographically broad (global) trade in them as well as an attendant environmental impact. Thus, bears in Montana, Idaho, and Alaska (and probably other states) are known to have been illegally killed to provide body parts for Asian folk remedies. Likewise, tiger preserves in India and Nepal are periodically raided and impacted by people who profit from medicinal demand. Underlying these threats to the natural world is global connectivity. If Alaska was an isolated place, or if trade and communications mechanisms did not connect Montana and China, then the bear population would be safe — at least as regards to that particular threat. For better or worse, however, bears and everything else are part of a world in which geographically local demand can effect worldwide resource supply, and have global environmental implications.

Migration and trade

Human beings are a species on the move. We migrate (see Chapter 12). We travel for business and leisure. We engage in global trade and commerce (see Chapter 15). And in the process we sometimes purposefully or inadvertently transfer a particular species from its current home to a part of the world where it was previously unknown, possibly with negative environmental consequences. Following are two examples.

Rabbits were unknown in Australia until 1859, when a dozen or so were purposefully introduced to serve as a source of food. Lacking natural predators in their new homeland, their numbers went ballistic, reaching about a billion within a century. This could rank as the ultimate bunny joke, except that sheep are important to the Australian economy, and five rabbits eat about as much grass as one sheep. Poisoning campaigns have subsequently killed off literally hundreds of millions of rabbits, but people now fret about the impact of those substances on local food chains (a concept discussed in the next section.)

In 1988, an East European freighter in the Great Lakes dumped some ballast water (whose sole purpose is to add weight to a ship that would otherwise bob like a cork), and with it probably introduced the zebra mussel to North America. This diminutive critter with cute little stripes is native to the Black and Caspian Seas, where certain duck species and crayfish keep its numbers in check. Lacking a large predator population in the Great Lakes — plus the tendency of adult female zebras mussels to produce more than 30,000 eggs per year — the population soared at a rate that would make an Australian rabbit swoon. The mussels attach to hard surfaces, and in the process have caked and stopped up water intake vents of waste and industrial treatment plants.

Focusing on food chains

While one may decry negative environmental impacts on ethical or scenic grounds, the repercussions for human well-being is perhaps of paramount importance. Food chains provide an excellent example of how negative impacts may come back to haunt humans big time. No, not *fast* food chains that provide hamburgers, tacos, fried chicken, pizza, or other staple foodstuffs; but instead the food chains that involve consuming and passing along injurious substances in ways that prove ultimately harmful or fatal to humans and animals.

The "chain" of events

A *food chain* is a sequence of living things through which energy and other matter move in an ecosystem. That mumbo-jumbo clearly calls for a diagram — two in fact, inasmuch as there are *terrestrial* (land) food chains and *aquatic* (water) ones (see Figure 18-3). Though their specific components differ, the overall structures are quite similar.

- At the base of any food chain are a number of *primary producers* that convert solar energy to organic matter — green plants in the case of terrestrial ecosystems, and aquatic plants or algae in the case of aquatic systems.

- Any animal that feeds upon a primary producer is a *primary consumer*. This may be a rabbit or squirrel (any herbivore will do) in the case of a terrestrial food chain, or a small fish in the case of an aquatic food chain.

- Any carnivore that feeds on a primary consumer is a *secondary consumer*.

- Any carnivore that feeds on a secondary consumer is called a *tertiary consumer,* and so forth. Humans can be primary, secondary, tertiary, or whatever consumers depending on what they are eating and whether that something had itself previously eaten another consumer.

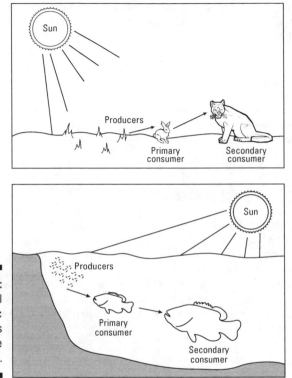

Figure 18-3:
Terrestrial
and aquatic
food chains
link diverse
life forms.

✔ All living things ultimately die (a major bummer) and decay.
Decomposers, which consist of various microorganisms and bacteria,
feed on organic waste and matter at all levels in the food chain and, as
their name suggests, aid in decomposition. Ultimately, they disintegrate
organic matter to an elemental chemical level that makes it suitable for
reuse by plants and animals — completing the food chain.

Thus, every food chain may be thought of as a set of linked organic units in
which matter is constantly used and recycled.

The potential for danger

While food chains sustain life in all its forms, they may also pose grave (perhaps fatal) harm when toxic substances are let loose in the environment. The
scenario may easily follow these steps:

1. **Toxic waste from a chemical manufacturer enters a lake and taints
   microscopic forms of life.**

2. **Small fish eat the tainted matter and store the harmful stuff in their
   fatty tissue.**

The role of attitudes toward nature

People generally have kindly attitudes toward some natural environments and less kindly attitudes toward others. These differences may result in concerted efforts to preserve some components of the natural world even as others are allowed or encouraged to be degraded or destroyed.

Consider, for example, *forest* and *swamp*. If you are like most people, then these two terms probably evoke different feelings. More specifically, most Americans have a rather benevolent and protective attitude towards forests and a much less benevolent and protective attitude toward swamps. Forests, of course, are places we humans can readily walk through and even make our home. Swamps, in contrast, are places we cannot similarly experience and are much less likely to call home. Wildlife may also play a role in explaining why these habitats are perceived so differently. Cartoons and other media have generated a generally positive mindset towards chipmunks, squirrels, deer and other forest animals (a phenomenon that has been called *the Bambi Complex*), but a less cuddly set of attitudes towards the reptiles and amphibians that tend to populate swamps.

This, of course, is bad news for swamps, which provide unique and necessary habitat for a great number of species. But how can one protect natural habitat and successfully lobby legislators to that effect if people simply have negative attitudes toward the thing that you seek to protect? In this case, the solution has been to drop *swamp* from the environmental vocabulary and replace it with *wetland*. Accordingly, a substantial body of environmental law aimed at wetland protection has been enacted during the past few decades and enjoys broad public support. People, it seems, are perfectly willing to protect a wetland rather than a swamp, even if the two places in question are one and the same.

3. **Bigger fish eat the smaller fish, and with each meal increase the concentration of toxic matter in their own fatty tissue, a process call *biological amplification*.**

4. **Fishermen catch the affected fish, eat them, and so acquire the toxic substance, perhaps with fatal result if enough contaminated fish are consumed.**

Fallout is another way in which hazardous materials can enter the environment. Fallout is a quaint term that most folks don't fully appreciate. When a big explosion occurs, a lot of solid matter is instantly pulverized and becomes very fine particles that are thrust into the air. Being so small, they can remain aloft for a long period until, because they have weight, they "fall out." Before that happens, however, wind may "spread the mess" over a wide area. If the explosion involves radioactive substances, then the fallout will be radioactive in nature. That alone may taint grass that is eaten by cattle and other livestock. Rainfall, however, may percolate radioactive matter into the soil, where it is taken up by roots and thus rather thoroughly contaminate the plants.

The Chernobyl disaster provides an instructive real-world case study. On April 26, 1986, a nuclear power plant in that Ukrainian (then Soviet) city suffered an explosion that resulted in discharge of a large quantity of radioactive matter into the atmosphere. Initial attempts by the government to squelch news of the incident were rendered futile when atmospheric monitoring stations in Sweden, hundreds of miles away, reported high levels of radioactive fallout.

In the case of Chernobyl, measurable amounts of fallout were recorded more than 1,200 miles away. As a result, the authorities ordered the destruction of thousands of cattle, pigs, and chickens out of fear that a contaminated food chain would lead to adverse human health impacts. Nevertheless, related sicknesses have subsequently occurred, and many public health officials predict more problems in the years ahead.

Going Global: Multiple Sources Affect an Entire Population

The Chernobyl and Exxon Valdez accidents discussed earlier in the chapter illustrate how a one-time point-source release of pollutants in a dynamic environment can have wide-ranging effects. More serious, however, is a case of on-going discharge from multiple-point sources, which generate potential for negative human impact on a global environmental scale. "Potential?" No, scratch that. At least two phenomena have already sprung forth: acid rain and global warming.

Raining acid

Acid rain is precipitation that has an unusually high acid content. It has been identified as a major cause of forest demise in various parts of the world, and also of contamination of lakes and estuaries.

The principal cause of this phenomenon is the burning of fossil fuels to operate motor vehicles, manufacturing facilities, and electrical utility plants (see Figure 18-4). A lesser but important cause is the smelting of ores that contain sulphur. Two unfortunate by-products are sulfur dioxide and nitrogen dioxide. When sulfur dioxide interacts with atmospheric moisture, it transforms into a dilute solution of sulfuric acid, a very corrosive substance. Nitrogen dioxide is transformed into nitric acid, which is not particularly corrosive, but is harmful to organic matter.

The range of impacts

After sulfur dioxide and nitrogen dioxide have formed in the atmosphere, they may remain there for long periods and be carried by the wind far from their source areas. Ultimately, they fall to earth with snow, sleet, hale, or rain. Even dew may bring them to Earth. Thus, though the phenomenon is popularly called acid rain, it would be more properly called *acid precipitation* or *acid deposition*. Whatever the label, negative impacts may be manifested in the following ways:

✔ **Terrestrial impacts:** Terrestrial impacts concern vegetation and soil. After acid precipitation has seeped into the ground, it may be taken up by the root systems of plants. If a sufficient amount of acid is taken in, even as small dosages over a long period, the stress may be sufficient to kill individual trees and perhaps destroy entire forests. Obviously, the effects of such outcomes may be devastating for commercial logging as well as wildlife habitat and recreation. Loss of vegetation results in decreases in the maze of root systems that hold soil in place, perhaps leading to erosion and mudslides. In addition, the acid may also have a fatal effect on bacteria and microbes that break down organic nutrients in the topsoil and contribute to soil fertility. Thus, acid rain may have a harmful effect on both the quality and quantity of soil.

✔ **Aquatic impacts:** Acid rain runoff may adversely impact streams, rivers, lakes, and oceans. Aquatic plants may be affected in the same manner as terrestrial ones. If they die, so does habitat for all kinds of animal species, as well as potentially significant elements in food chains. In addition, fish and other animal species may perish directly from prolonged exposure to the acids, or indirectly from silt laden runoff attendant to loss of roots that held soil in place.

Small lakes that collect runoff, but have no outlet, are particularly susceptible to harm. Literally and figuratively, they are dead ends, the final collecting place for acid rain. Acid concentrations may be such that lakes become, for all intents and purposes, sterile.

Estuaries may also be impacted, and with them, marine fisheries. The problem here is not so much the acid as the nitrates, a favorite of algae. The impact of that, as was discussed earlier in the chapter in the case of the Chesapeake Bay, may be fish kills on a significant scale.

✔ **Material impacts:** Material impacts principally concern deterioration of stonework and *statuary* (sculptures and statues), particularly ones made of marble. The corrosive effect of acid rain may literally eat away at these to the point where — to use an admittedly extreme example — statues become undistinguishable slabs of rock. Building facades are of more immediate importance, and thousands of them around the world are slowly dissolving.

Applied Geography: Watershed management

A *watershed* is an area that drains into a river, lake, bay, or reservoir. As we saw with the case of the Chesapeake Bay, the quality of a body of water may be intimately related to the quality of the watershed that feeds it. Watershed management concerns a wide rage of efforts that monitor and ensure the quality of watersheds and, thusly, the receiving bodies. This is a significant environmental undertaking in any case, but especially so when municipal or regional drinking water is at stake.

Various geographical techniques are routinely applied to this endeavor. The Global Positioning System (see Chapter 4), for example, is used in field surveys to pinpoint and map pollution sources. Aerial photography and satellite imagery may be used to assess the health of vegetation that anchors watershed soils. But perhaps the most intriguing technology is *digital terrain modeling,* in which the lay of the land in exacting detail is stored in computer memory, allowing managers to simulate all kinds of eventualities. Want to see what will happen if a small pollution source becomes a big one? Or if acid rain should reduce forest cover by 55 percent (or any other percentage)? Or if 2 inches of rain should fall anywhere or everywhere in 25 minutes? This computer technology facilitates simulation of all kinds of scenarios, the goal being watershed management that promotes and preserves the quality of drinking water.

Figure 18-4: The acid rain that destroys forests and contaminates lakes largely originates in personal, commercial, and industrial consumption of fossil fuels.

Acid rain

Sulfur dioxide
Nitrogen oxides

factory

Car

Semi-truck

Acids flow into lake and kill fish

Tree roots take in acids

The geographical dimension

To some extent, acid rain happens everywhere, but its intensity is very uneven (Figure 18-5). In general, the world's principal industrial regions tend to the major source areas. That includes the southeastern Great Lakes region, Western Europe in general, plus the Ukraine and southwestern Russia,

Eastern China, and Japan. Acid rain tends to be greatest in these immediate regions plus those directly downwind. Given the general west-to-east movement of the atmosphere at mid latitudes, the typical result is an elongated impact area that tails off easterly from the source region.

Thus, in North America, relatively pristine areas in northern New York, New England, and Canada's Maritime Provinces have been hard hit. Forests are dying over large areas and numerous lakes, especially small ones, are becoming lifeless. The situation is similar in Europe, where the Black Forest and similar tracts are being affected.

Sometimes a state or country finds itself on the receiving end of pollution that originates in another state or country. This is called *trans-boundary pollution*. Thus, the nitrates that cause algae blooms that kill fish and hurt the economy in Maryland's portion of the Chesapeake Bay may have their origin in fertilizers that were spread on agricultural fields in Pennsylvania. Similarly, acid rain that kills forests in eastern Canada may have their origins in power plants in the United States.

Figure 18-5:
While the effects of acid rain are global, the greatest impacts are felt in regions downwind from major source areas.

Source Areas of Acid Deposition
Source Areas and Problem Areas

People who live in areas that suffer the effects of pollution (such as acid rain) tend to be among the most committed to finding a remedy. But how can you affect a solution to a problem that originates outside of your jurisdiction — in another state or country? Assuming no treaties or protocols have been broken, about all one can do is encourage your neighbors to clean up their act. In the case of acid rain, that is going to take some doing because so much of the problem is associated with fossil fuel consumption that is likely to continue for decades to come.

Getting warmer

Global warming is the term used to describe the increase in the average temperature of the atmosphere that has been occurring during the past several decades. This has ominous implications for sea level rise and local climate changes of various sorts. But why is global warming happening? Is it a natural process or the result of human activity? And can the process be slowed or reversed?

The Principal suspects

Several ice ages came and went long before *Homo sapiens* emerged as a major instrument of global change. That means that natural mechanisms capable of effecting significant global temperature change over time must exist. Perhaps, therefore, the present warming trend is merely part of a natural cycle that has occurred in the past, is happening at present, and is likely to repeat in the future. Or perhaps it is the result of some other natural cause that we have yet to identify.

On the other hand, the warming trend may be the result of atmospheric changes for which humans are responsible. Since the beginning of the Industrial Revolution in the mid- to late- 1700's, humans have consumed (burned) large quantities of coal, petroleum, and natural gas to operate vehicles and machinery, provide power for manufacturing facilities and electrical utilities, and to heat up and cool down homes and buildings. In the process, huge quantities of certain gases have been added to the atmosphere. Most reputable scientists believe that these substances are the cause of the current warming trend.

The global greenhouse

The commonly held scenario for why the atmosphere is heating up likens the atmosphere to a greenhouse, whose glass panes allow sunshine to penetrate the structure, but then contain the solar energy's heat and prevent it from escaping. Of course, no giant glass roof hovers over planet Earth. But gases in the atmosphere do act in much the same way (Figure 18-6). When these gases are scarce, a considerable amount of solar energy is able to radiate off Earth's surface back into the coldness of space. When they are abundant, however, they produce an atmosphere that absorbs or "traps" this same heat, resulting in a warming atmosphere. The gases that accomplish this are called *greenhouse gases* and the over-all process is referred to as the *greenhouse effect*.

Following are two of the primary contributors to global warming:

- **Effects of the Industrial Revolution:** Carbon dioxide (CO_2) is the most prolific greenhouse gas. Although it's a naturally occurring substance, substantial quantities of it are released to the atmosphere when fossil fuels, wood, and other organic matter are burned. Increased fuel consumption, especially as regards to fossil fuels, has been a hallmark of the

Industrial Revolution. Before it began, the carbon dioxide content of the atmosphere measured about 274 parts per million (ppm). By the beginning of the present century, that number had risen to more than 360 ppm. Because global consumption of fossil fuels continues to rise, one can only expect increasing levels of greenhouse gases in the atmosphere, and with them the likelihood of increasing temperatures.

Emissions vary geographically. Because economic development and fuel consumption go hand-in-hand, understandably developed nations generate a much greater portion of annual emissions than do developing ones. And because the United States is the world leader in fossil fuel consumption (see Chapter 16), it makes sense that it also leads the world in greenhouse gas emission. During the 20th Century, the U.S. was responsible for an estimated 30.3 percent of global carbon dioxide emissions. Europe accounted for another 27.7 percent and the lands of the former Soviet Union another 13.7 percent

✔ **Effects of deforestation:** This era of increased fuel consumption has been complemented by a general process of deforestation — first in Eurasia, then in North America, and now in the tropical rainforest areas. Trees take in carbon dioxide from the atmosphere and give off oxygen. Thus, when trees are cut and consumed, they are no longer available to effect this exchange. And if they are burned, they emit the CO_2 stored in them, increasing the atmospheric store of that gas. Experts disagree about the impact of the current decline of tropical rainforest on global atmospheric balance, but the majority opinion is that its contribution to the greenhouse equation is measurable, if not significant.

Figure 18-6:
Large quantities of carbon dioxide in the atmosphere greatly contribute to the global greenhouse effect.

Winners and losers

If you live in an area that experiences a cold winter, then you may be excused for thinking that global warming can't be all bad. If continued increase in global warming comes to pass, then clearly some lands will benefit. For example, large parts of Canada and Siberia that are now too cold for agriculture are likely to become productive land.

But mathematical models predict that large, productive areas in the middle latitudes will dry up and become desert. Clearly, sea level will rise, resulting in coastal inundation and an overall decrease in land available for settlement. And with that will come salt-water intrusion of neighboring aquifers, resulting in an overall decline in the global supply of potable water. For these sorts of reasons, most models forecast more losers than winners should the warming trend continue. And that is likely given continued and growing consumption of fossil fuels.

Taking on the Challenges of Tomorrow

This chapter began by describing environmental events in Chesapeake Bay in terms of its connections to a much broader geographical expanse. In a manner of speaking, the same applies to you. Wherever you live, you are part of an environment that, by virtue of dynamic mechanisms, is connected to other places that may be miles away or next-door. While the results may affect you, you also have the capacity to effect results. Specifically, as citizens of a democracy, you have the opportunity to participate in decision-making that can effect at varying scales the quality of the world in which you live. Following are two examples that U.S. citizens have the opportunity to become geographically involved in.

ANWR

Some 19 million barrels of petroleum are consumed in the United States each and every day, and more than half of them are imported. That kind of consumption scenario — particularly the reliance on foreign sources of oil — makes a strong argument for tapping domestic oil reserves to the fullest. But what if an area of proven potential coincides with a pristine wildlife reserve? Americans love their petroleum, but they also champion wildlife and wilderness. Can we have it both ways? Is drilling compatible with wildlife? Should it be prohibited in certain areas? Or should our need for oil dictate that we go ahead and drill wherever we can as long as we make an honest effort to protect the environment as best we can?

These questions are at the center of the debate concerning the Arctic National Wildlife Refuge (ANWR), a South Carolina-sized area in Alaska's extreme northeast corner (Figure 18-7). The same Congressional Act that created the Refuge's present boundary also designated 1.5 million acres of it along the coastal plain as a potential future site for oil exploration. That area is now estimated to contain anywhere between 4 and 16 billion barrels of oil (depending on whose data you consult), although the amount that can be profitably recovered is probably no more than 6 billion barrels. That figure turns out to be less than a year's worth of annual U.S. petroleum consumption.

The zone with the greatest oil-producing potential lies within the principal calving area of the porcupine caribou, whose annual migration brings a herd of some 130,000 members to the coastal plain each summer. Numerous other species frequent the area for all or part of the year — after all, it is a National Wildlife Refuge. Debate rages concerning the effects of drilling (together with attendant construction and road building) on wildlife and whether or not the productive lifetime of the oil reserve (perhaps 40 years) is worth the risk. Ultimately, your elected representatives will make the decision.

Figure 18-7:
The Arctic National Wildlife Reserve in Alaska.

Garbage and NIMBY

At the end of the day you take out the garbage. But where does it go? Or where should it go? A popular refrain is NIMBY — not in my back yard. "Put it somewhere, but not next to me."

Refuse is very geographical on at least two accounts. First, a direct relationship exists between the geography of wealth and the geography of garbage.

That is, the more affluent the society, the greater the amount and durability of its garbage. Because the United States is the world's most developed country, it should come as no surprise that it ranks as the world leader in garbage production, whether measured nationally or per capita.

Second, garbage has got to go somewhere. But where? Landfills and incineration are the most popular responses, but both are susceptible to dynamic mechanisms — percolating water in the case of landfills, and wind as regards incinerators. Nobody wants it in their backyard, but increasingly the cities and counties that generate it are being forced to dispose of it somewhere within their own jurisdictions, as opposed, say, to empty lands far away. And once again, the question of "Where?" will be answered by public officials that one hopes are elected by and responsive to a geographically savvy citizenry.

Part V
The Part of Tens

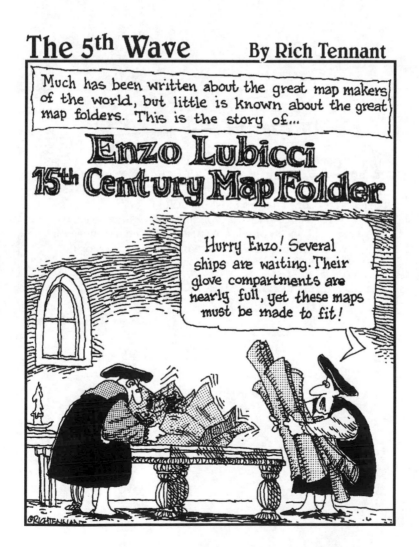

The 5th Wave   By Rich Tennant

Much has been written about the great map makers of the world, but little is known about the great map folders. This is the story of...

Enzo Lubicci 15th Century Map Folder

Hurry Enzo! Several ships are waiting. Their glove compartments are nearly full, yet these maps must be made to fit!

In this part . . .

Lists sure are popular these days. The 20 best so and so. The 15 worst such and such. Ten, however, seems to be the most popular number. As in "The Ten Best Dressed Celebrities," or "The Ten Most Beautiful Beaches." In this part, you will encounter 10 geographical organizations, 10 geographical occupations, and 10 geographical things you can forget. The first two lists are kinda serious. As for the last one, well. . . .

Chapter 19

Ten Organizations for Geographic Information

A number of public and private organizations provide information, services and products of interest. Here are ten of the most noteworthy.

Aerial Photography Field Office (APFO), United States Department of Agriculture

How would you like to acquire aerial photographs of, say, your home town as it appeared in a recent year, as well as during the 1980s, the 1970s, the 1960s, and so forth? Aerial photographs are great tools for analyzing the geography of an area and how it has changed over time. And when it comes to air photos of the U.S., nobody beats APFO. This branch of the Department of Agriculture serves as the storehouse for aerial photographs of the United States produced under government backing from as far back as the 1930s. The inventory consists of literally millions of air photos, all of which are available to the public. Prices start at $6 for 9" x 9" black-and-white prints that were standard fare through the mid-1970s, and go higher for infrared prints from more recent aerial surveys. The costs are a fraction of what you would pay a local aerial survey firm for the very same thing, the difference being that APFO is a non-profit government agency.

You can visit the agency online at www.apfo.usda.gov, or write them at USDA-APFO, 2222 West 2300 South, Salt Lake City, UT 84119-2020. Their phone number is (801) 975-3503.

American Congress on Surveying and Mapping (ACSM)

The ACSM is a non-profit educational organization made up of four member societies which serve as special interest groups: The American Association of Geodetic Surveying; The Cartography and Geographic Information Society; The Geographic and Land Information Society; and The National Society of Professional Surveyors, Inc. Its objectives are to advance the sciences of surveying, mapping, and related fields; to encourage the development of educational programs in surveying, mapping, and charting; and to support publications that represent the professional and technical interests of surveying and mapping.

This is a professional organization, so novices may feel a bit overwhelmed. However, if you are seriously interested in mapmaking, learning more about cartographic technology, or finding career opportunities and preparation in map-related fields, then familiarity with ACSM is recommended. Additional information about the organization and its constituent bodies may be obtained online at www.acsm.net. You can also reach them the old-fashioned way at American Congress on Surveying and Mapping, 6 Montgomery Village Avenue, Suite #403, Gaithersburg, MD 20879. Their phone number is (240) 632-9716, ext. 108.

Association of American Geographers (AAG)

The AAG is the pre-eminent organization of professional geographers in the United States. The organization's goals are to promote professional studies in geography and to encourage the application of geography in business, education, and government. The AAG publishes two scholarly journals (*The Annals of the Association of American Geographers* and *The Professional Geographer*), a monthly newsletter, and annually hosts a national convention and several regional meetings. Membership is not recommended for the average novice, but if you are interested in furthering your geographic education at the university level, then the AAG should be kept in mind.

The AAG maintains an informative Web site (www.aag.org) concerning the organization and geography in general. When you access their homepage, I particularly encourage you to click on "Careers in Geography" and browse the results. Any questions about any aspect of the AAG or geography in general? Click on "Ask a Geographer" and start typing. Don't have a means of accessing the Internet? Write 'em a letter at AAG, 1710 Sixteenth St. NW, Washington, DC 20009-3198, or give them a ring at (202) 234-1450. Of course, you can always fax them a letter at (202) 234-2744.

National Council for Geographic Education (NCGE)

The NCGE seeks to enhance the status and quality of geography education and learning. Familiarization with this organization is recommended for anyone interested in geographic education at any level, kindergarten through graduate school. The *Journal of Geography,* the NCGE's principal periodical, is published ten times a year and is devoted to research on geographic education, content articles, and classroom-ready lesson plans and activities. The Council's newsletter, *Perspective,* published six times per year, provides information on learning activities, current events, educational resources, workshops, and other items related to the teaching of geography, as well as to NCGE news. The organization holds an annual meeting, the venue of which changes each year, and has available for sale a variety of monographs related to its mission.

You can learn more about this organization online at www.ncge.org or by sending an e-mail to ncge-org@grove.iup.edu. The NCGE's Central Office is located at Indiana University of Pennsylvania, Leonard Hall Room 16 A, 421 North Walk, Indiana, PA 15705-1087. Feel free to call them at (724) 357-6290 or fax them at (724) 357-7708. (Note: Effective June 30, 2002, the Central Office will relocate to 206 A Martin Hall, Jacksonville State University, Jacksonville, AL 36265. Phone: [800] 346-5444.)

National Geographic Society (NGS)

National Geographic is more than just a magazine. Founded in 1888 "for the increase and diffusion of geographic knowledge," the Society is by far the most visible symbol and promoter of popular geography in the United States. In recent years, the Society's reach and influence have been augmented by its *National Geographic Explorer* TV series, a cable TV channel, its own magazine of travel and tourism, *National Geographic Traveler,* and another publication geared to the more active reader called *National Geographic Adventure.* Meanwhile the old stalwart, *National Geographic Magazine,* continues to rank among the top periodicals in terms of circulation, and has recently morphed into foreign language versions.

The Society has become a force in geographic education, as evidenced by its pivotal role in producing and promoting the National Geography Standards, its sponsorship of the National Geography Bee, and its nurturing of the state Geographic Alliance Network (see the last item on this list). The Society is also a major marketer of maps, globes, atlases, and assorted geographical paraphernalia. Information may be accessed online at `www.nationalgeographic.com`, which also offers a ton of downloadable goodies. You can also call (800) NGS-LINE, or write to National Geographic Society, 1145 17th St., NW, Washington, DC 20036-4688.

National Oceanic and Atmospheric Administration (NOAA)

NOAA is the principal agency of the U.S. government concerned with collection, analysis, and dissemination of information about the ocean and atmosphere. As such, it offers a wealth of data regarding non-terrestrial aspects of physical geography. Its oceanic concerns involve coral reefs, tides and currents, marine sanctuaries, coastal development issues, and the effects of chemical and oil spills. Given its mandates regarding safe navigation and transportation, NOAA also engages in ocean mapping and is the country's major publisher and supplier of nautical and aeronautical maps and charts.

Atmospherically, NOAA is probably best known as the parent agency of the U.S. Weather Service, whose official forecasts, advisories, and satellite images are standard fare for TV and radio weather reports. Climatic phenomena, such as El Niño and La Niña, global warming, climate prediction, and *paleoclimatology* (ancient climate patterns) fall within its scope, as well as topics about atmospheric quality and human-atmosphere interaction.

A visit to NOAA's well-organized Web site is a great way to explore what the agency has to offer and to obtain hard-copies of their information. The online address is `www.noaa.gov`. You can also reach them through their Public Affairs Office at (202) 482-6090, or write to NOAA, 14th Street and Constitution Avenue, NW, HCHB 6013, Washington, DC 20230.

Population Reference Bureau (PRB)

The Population Reference Bureau provides timely information on population trends and their implications. While it markets some very fine resources pertaining to American demographics, PRB's signature product is its *World Population Data Sheet* ($6, including postage and handling), an annually updated chart that includes more than 15 demographic variables for

200 countries. Other publications include the quarterly *Population Bulletin,* a monthly newsletter (*Population Today*), and the quarterly *PRB Reports on America.* The PRB also has an education program that produces high-quality lesson plans and classroom resources for the elementary and secondary school levels.

PRB is completely apolitical. It does not endorse or advocate positions related to population policy or controversial population issues. In short, the PRB has no agenda other than to provide the most reliable information possible, and to let the data speak for themselves. Access their Web site at `www.prb.org` or send them an e-mail at `popref@prb.org`. You can snail-mail them at PRB, Suite 520, 1875 Connecticut Avenue, NW, Washington, DC 20009-5728. You can always try one of their phone numbers at (202) 483-1100 or (800) 877-9881. If all else fails, send them a fax at (202) 328-3937.

United States Census Bureau

This is the place to go for information on U.S. (and, to a lesser extent, global) population geography. The census happens only once a decade, but the U.S. Census Bureau is a perennial organization with a lot to offer. Naturally, most of that is data derived from the census and organized by various geographic levels — national, state, county, city, census-tract (rather like a postal zip code), and occasionally city blocks. But the bureau is not just demographics. It provides data on housing, businesses of all sorts, foreign trade, genealogy, and other matters. Educational resources for teachers also are available.

Part of the Bureau's mission is to make its data readily available to the general public. Accordingly, reams of information can be accessed and downloaded from the agency's Web site, `www.census.gov`, which also provides links to more than a hundred counterparts and related agencies around the world. For general information, you can call the Bureau at (301) 457-4608, or write to U.S. Census Bureau, Washington, DC 20233.

United States Geological Survey (USGS)

Created by an Act of Congress in 1879, the USGS is the major science agency of the federal government concerned with our physical planet and its Earthly processes. While the name rather matter-of-factly recounts the agency's original mission, its scope has expanded to include ecosystems, endangered species, environmental quality, and natural hazards of all sorts. The USGS is also renowned for its handsome and exquisitely detailed topographic maps of the United States, which are available to the public.

Like many other government agencies, the USGS has a public outreach mandate. And nowhere is this more evident than in its robust and well-organized Web site, www.usgs.gov. Want current information on earthquakes, volcanoes, floods and wildfires? Click on CINDI, the Center for Integration of Natural Disaster Information. Want the latest assessment on global petroleum reserves, the impact of oil drilling on the Arctic National Wildlife Preserve, or the status of West Nile virus? Click on USGS headlines. General contact information is as follows: USGS National Center, 12201 Sunrise Valley Drive, Reston, VA 20192. Or try calling at (703) 648-4000 or (888) ASK-USGS.

Your State's Geographic Alliance

If you are interested in geography education in your state — particularly at the K-12 level — then you should become familiar with your state's geographic alliance. I refer to the Alabama Geographic Alliance, the Alaska Geographic Alliance, the Arkansas Geographic Alliance, and so forth. One can be found in every state, plus Puerto Rico, Canada, and the Dominican Republic. The alliance network is a product of The National Geographic Society's Geography Education Foundation, created in the late 1980s to help resuscitate the teaching of geography in the nation's schools. The Alliances sponsor and conduct workshops and teacher-training institutes, prepare and disseminate lesson plans and other education materials, and publish newsletters — all focusing on geography as it pertains to specific curricula in the individual states. In most cases, alliance membership is free and is the key to getting on the newsletter mailing list and finding out what's happening in your state. A list of appropriate names and addresses is available on the National Geographic Society's Web site. Go to www.nationalgeographic.com/education/teacher_community/, click on "Select your state" and do just that.

Chapter 20

Ten Interesting Geographical Occupations

*H*undreds of geography departments in colleges and universities around the world offer courses and degree programs that prepare people for exciting and rewarding careers. Here are ten examples of how geographic education gets put to work.

Air Photo Interpreter/Remote Sensing Analyst

Aerial photographs and satellite images are key sources of information about Earth's surface and, therefore, important tools for geographic analysis. Numerous government agencies employ thousands of interpreters and analysts to monitor goings on in foreign countries as well as to keep tabs on domestic agriculture, forestry, and other items of environmental interest. City and regional planning authorities, as well as civil engineering and consulting firms, also hire specialists who use these tools to acquire information about their employers' respective areas of interest and provide input into maps and databases that aid planning. Indeed, nearly all maps produced under government auspices these days are based on information provided by air photos and satellite images.

Area Specialist

An area specialist is a person who possesses a high degree of expertise in a particular part of the world. Latin Americanist, Africanist, and Sinologist (China expert) are typical job titles. Government agencies and international business concerns are the principal employers. Virtually hundreds of such people are hired by the U.S. State Department and Central Intelligence Agency to "keep current" on their region (mainly by reading newspapers and government reports printed in the region of expertise) so as to be able to provide the best possible information and advise to policy makers. In the business world, area specialists act in much the same capacity, focusing particularly on economic geography, human/cultural geography, and climatology. As the global economy becomes more interconnected, job opportunities for area specialists in the business world are likely to grow.

Cartographer/GIS Specialist

Cartographers and Geographic Information Systems (GIS) specialists are concerned with production of maps, which are indispensable tools for displaying information about places and regions. Government agencies of all kinds use them for analytical and display purposes — ditto for construction and utility companies, architectural and engineering firms, and a host of other employers. The U.S. Defense Mapping Agency alone has thousands of cartographers in its hire, and that is just the tip of an employment iceberg that keeps growing. Nowadays, professional cartography is largely computer-based and complemented by GIS, which facilitates storage and mapping of a wide range of information about places.

Educator

The prospects for landing a job as a geography teacher in this country are the best they have been in at least 30 years and keep getting stronger. Sure, a long time ago geography largely disappeared from the curriculum as a stand-alone subject and became submerged in that academic goulash called social studies. Times have changed. The National Standards movement has not only resuscitated the teaching of geography, but has also mandated in many states that those who teach it have solid academic preparation in the subject. Sadly, the number of people who are so qualified is rather slim. The positive flip side of that statement is that opportunity awaits tomorrow's teachers who specialize or major in geography. Also, growing awareness of the importance of geographic perspectives and technology by the government and business communities bodes well for future prospects in college teaching.

Environmental Manager

Environmental managers monitor and protect natural resources. Environmental protection agencies hire these professionals, as do public and private enterprises involved in waste disposal, water supply, forest conservation, wetland and coastal zone management, and other resource-related pursuits. Monitoring compliance with environmental laws is a major concern of environmental managers, as are preparation and review of environmental impact statements. Because environmental phenomena characterize different parts of Earth's surface, they are an important field of study for geographers. Geography programs in colleges and universities offer a range of environmentally related coursework. Also, expertise in cartography, geographic information systems, air photo interpretation, remote sensing, and/or spatial data analysis greatly enhances career prospects in this field.

Health Services Planner

These planners seek to optimize delivery of and access to health services. Some of these professionals analyze and recommend locations for clinics, hospitals, and emergency response units. Others help to decide how to allocate specialized services (such as coronary care and burn units) among existing hospitals and health facilities. Still others analyze patterns of disease and health risks with a view to geographically targeting outreach and education programs aimed at prevention. While most of these planners have degrees in public health, many also turn to geography programs to acquire critical skills in cartography, geographic information systems, and spatial statistics.

Location Analyst

An old proverb says that the three most important determinants of the success of a business are location, location, and location. Although this is not literally true for every kind of business, the right location is often the difference between success and failure. Location analysts identify the factors that most affect (for better or worse) the success of a particular business, evaluate the viability of prospective sites with respect to those factors, and recommend a site to decision-makers. Many large (and not-so-large) retail chains maintain in-house think tanks of location analysts to help make all-important decisions regarding the locations of future stores. Numerous generic consulting firms do the same. Location analysis is an important component of economic geography courses.

Market Analyst

No, not the stock market, but instead, the prospective buyers of particular goods and services. People have different tendencies to purchase different things. Such variation may be a function of age, culture, race, ethnicity, income, or some other factor that may vary geographically. Determining the geography of the market for particular products and allocating merchandise accordingly is the job of the market analyst. You think this isn't important? Some time ago, a nationwide retailer (that shall remain nameless) sold decent merchandise, but nevertheless went belly up, in part because of poor market analysis. Thus, they would open a 10,000 square foot store in a city that could easily support a 50,000 square foot store, while opening a 50,000 square foot store in a town that could only support a 10,000 square foot store. In other instances, they would allocate significant floor space to maternity clothing, baby goods, and toys in towns with a high percentage of retirees. Mind you, their location analysis was great. They had nice stores in superb locations. They were just the wrong-sized stores (in many instances) with the wrong product mix in the wrong locations. Good market analysis, a staple subject of economic geography, could have made a difference.

Transportation Planner

Transportation concerns movement of goods and people from one location to another. Thus, transportation planning is inherently geographical and attracts people with training in geography. Several specializations are involved. Surveyors who plan for future road construction typically are well-grounded in modern cartographic skills. Route planners concern themselves with devising optimal itineraries for public buses and planning for evacuation of areas in the face of diverse emergencies. Meanwhile, in the realm of manufacturing, appreciation is growing for "just in time" planning, which seeks to bring together product components at a factory . . . well, just in time for assembly, thus reducing significant warehousing costs. All of this rightly implies that the job outlook for transportation planning specialists is excellent.

Urban Planner

Cities consist of countless land parcels devoted to residential, commercial, industrial, cultural, recreational, transportation, governmental, sanitation, and other uses. Urban planners geographically allocate the various kinds of parcels in ways that promote orderly development of cities and enhance their attractiveness as places to live and work. Virtually thousands of counties and municipalities in the United States employ urban planners. Governments of nearly all major cities include a department of city planning that employs dozens of specialists in the field. The fact that cities keep growing and that 75 percent of Americans now live in them bodes well for the future of urban planning as a career field. Cartography, geographic information systems, air photo interpretation, and analysis of spatial databases have emerged as critical tools of the planning trade.

Chapter 21

Ten Things You Can Forget

*I*f reading about all of this geography stuff has left your brain low on disk space, then you may wish to delete some files and free up some memory. Accordingly, here are ten geography-related items you can relegate to the recycle bin.

The Bermuda Triangle

The Bermuda Triangle is an imaginary area off the southeastern coast of the United States that has allegedly witnessed a disproportionately high number of strange and unexplained disappearances of boats, ships, and airplanes. Disagreement abounds as to its exact location. Most commonly, the Triangle is depicted as a hunk of Atlantic Ocean bounded by a line that extends from Bermuda to Miami to San Juan, Puerto Rico and back to Bermuda. The U.S. Board of Geographic Names does not recognize "Bermuda Triangle" and has never indicated where it may be. This gives you complete license to place the Bermuda Triangle anywhere your little heart desires, and that is precisely what some people have done. Indeed, on some maps the Triangle isn't even triangular.

The term first appeared in a February 1964 article ("The Deadly Bermuda Triangle") in *Argosy* magazine. Several other articles on the subject subsequently appeared (along with books and TV documentaries), many suggesting the region is mysterious and dangerous. If you've seen some of this stuff and found the whole thing rather spooky, rest assured you are not alone. Given widespread public interest in paranormal activity, the Bermuda Triangle has become something of a geographical poster child for enthusiasts of the occult.

In reality, the Bermuda Triangle is much ado about nothing. The vast majority of the supposedly abnormal disappearances have very normal explanations, and some of the most famous incidents occurred well outside the Triangle. Several years ago, a researcher compiled an extensive list of sinkings and crashes, based on Lloyds of London accident reports and similar reliable sources, and plotted them on a map. If anything, the data show that the Triangle is remarkable for its lack of mishaps and is practically the safest part of the western Atlantic Ocean.

Cold Canadian Air

It happens several times each winter. I'm watching a weather report on TV and the map indicates frigid temperatures up in parts of Canada, such as Alberta, Saskatchewan, and Manitoba. With squinting eyes and stern gaze, the usually smiling weather-person glumly warns that "cold Canadian air" is on the way. Cold Canadian air — have you ever heard that term? If you live in the U.S., pay extra special attention to weather forecasts between November and March. Sooner or later you're going to be told that cold Canadian air is on the way. I guarantee it.

Given the imminent arrival of cold "Canadian air," it seems to me that you have every right to say "Darn those Canadians! Why don't they keep their cold air to themselves?" Of course, "Canadian air" does not exist. Neither does German air, Pakistani air, Ethiopian air, etc. Air is air. It doesn't belong to anybody. "Air space," the atmospheric area that is vertically overhead a country, is another matter. That belongs to Canada, according to international law. Thus, aircraft of foreign countries are not supposed to fly over Canada without the Canadians' permission. But Canada owns the physical space, not the air itself.

If the air over Canada is very, very cold, rest assured the Canadian people had nothing to do with it. And if that same cold air is coming your way, again rest assured that there is no nefarious neighbor to your north who is responsible. If you feel like blaming someone, or something, then consult Chapter 9, wherein you will come across a couple of environmental factors. Cold "Canadian Air?" Forget about it.

"The Rain in Spain Stays Mainly on the Plain"

I hate to be the one to tell you this, but "Wrong! Wrong! Wrong!" The plain (or *meseta*) that occupies central Spain is among the driest parts of the country. Madrid gets about 17 inches of rainfall on average per year. Valladolid, a city in another part of the plain, averages a little less than 15 inches. But it's another story in Galicia, the mountainous area in extreme northwestern Spain. Santiago de Compostela, the world-famous Galician center of pilgrimage, averages about 56 inches of rain per year. As moisture-laden air enters Galicia from the Atlantic, it rises to cross the mountains, and in doing so cools, condenses, and forms rain. By the time this air reaches the central plain, it's low on moisture, so rain is relatively scarce. I know it makes for poor poetry, but "The rain in Spain stays mainly on the mountains in Galicia." You don't have to memorize that if you do not want to. But as regards to "The Rain in Spain Stays Mainly on the Plain", unless you're preparing for a role in your town's production of *My Fair Lady,* forget about it.

"Coming Out of Nowhere"

Watch a football game on TV and sooner or later somebody will "come out of nowhere." Typically, a quarterback drops back to pass, spots a wide receiver in the open far downfield, and heaves the ball. It looks like a sure completion. But at the last millisecond, the cornerback or safety lunges to deflect the pass and save the day. The crowd screams and so does the play-by-play announcer: "Wow! He came out of nowhere to make that play!" It seems athletes have this knack for "coming out of nowhere" to do something spectacular. And it's not just football. Hockey, baseball, and basketball players do it, too. Usually, it involves a great play on defense, but not always.

In the years before instant replay, I wondered how somebody could be nowhere, yet come out of that non-location to perform a feat of great athleticism. It was one of the great mysteries of geography. I figured, maybe the person coming out of nowhere was in one of those parallel universes they keep talking about on *Star Trek* reruns. But then along came instant replay and guess what? That defensive back was there all along. So there was no great mystery of geography after all. But that did create a new mystery — why do TV networks give fat contracts to blind play-by-play announcers? I'm still searching for an answer to that question.

Land of the Midnight Sun

Alaska is called the "Land of the Midnight Sun." The future forty-ninth state became a U.S. territory in 1867 when, in the greatest real estate deal since Jefferson purchased the Louisiana Territory from Napoleon, Secretary of State William H. Seward negotiated its purchase from the Russian government for $7.2 million. People thought Alaska was worthless and Seward was nuts, so they called Alaska "Seward's Folly" and "Seward's Ice Box." Either might have become the moniker, so you got to admit "Land of the Midnight Sun" sounds pretty good. But as far as truth-in-geography is considered, that's another story.

For part of the year the sun does indeed shine at midnight north of the Arctic Circle. But only about 25 percent of Alaska is within that realm, and likewise only about 1 percent of Alaskans. Thus, if you want to get picky about it (and of course I do), the State's nickname does not apply to 75 percent of its land and 99 percent of its people. Certainly, nights are very short over most of Alaska during the summer. Indeed, nighttime may be more like a prolonged twilight. But midnight sun? Sorry. That only works north of the Arctic Circle. And there's one more detail that has been conveniently overlooked. Any locale that experiences midnight sun at one time of the year will also experience noon-time darkness at another time of year (see Chapter 7). "Land of the Noon-time Darkness?" Put that on your license plate! Actually, yes, I do get the point of it all. "Land of the Midnight Sun" is about image and tourism and putting your best foot forward. But as far as truth-in-geography is concerned, forget about it.

Tropical Paradise

Tropical + paradise. They go together like rocket + science, or banana + split. The genesis of this verbal union is, of course, The Book of Genesis from the *Bible*. Adam and Eve wore no clothes, so the Garden of Eden must have been warm. And since the Garden was a garden, it must have had a fair amount of rain. Warm + rain = tropical. Therefore, Paradise (capital P) was tropical.

Until fairly recent times, lots of people believed the biblical Paradise was real. That meant not only that the Bible was literally true concerning the Garden of Eden, but that it actually existed and was awaiting rediscovery. Some medieval world maps even showed an island off East Asia named Paradise. So when the early Spanish voyagers, thinking they had reached East Asia, encountered lush Caribbean islands, they truly believed they had found the tropical Paradise. Today, of course, "tropical paradise" has much more to do with tourism than theology. Typically, it appears in ads that encourage you to dispose of your disposable income by vacationing in some tropical destination.

It turns out, however, that just about every tropical paradise is a developing country with all of the social and economic problems that come with that title. Accordingly, many tropical resorts are a kind of compound that purposefully insulates affluent visitors from local realities. Please understand that this is not a put-down of those locales and certainly not of the people who live there. The residents tend to be honest, decent, hard-working, peace-loving people who are proud of where they come from. And as well they should. But with all due respect, just delete "tropical paradise" from your memory.

The Democratic Republic of . . .

Every country has an official name, and it usually differs from the name by which it is commonly called. Thus, Ireland is really The Republic of Ireland, and Thailand is really the Kingdom of Thailand, and so forth. Some of the "real" names are wonderfully evocative. Examples include The Hashemite Kingdom of Jordan and The Grand Duchy of Luxembourg. Great names! But then there are other official names that I look at and think, "You gotta be kidding me." Examples include The Democratic People's Republic of Korea (commonly called North Korea), The Lao People's Democratic Republic (Laos), and the People's Democratic Republic of Ethiopia. Each of these countries is currently under a virtual dictatorship. In fact, what "democratic republics" seem to have in common is a lack of characteristics that most people associate with democracy,like periodic honest elections, secret ballots, and two or more honest-to-goodness political parties. Are we talking bona fide countries? Yep. But should they be called "Democratic Republics"? Let's take a vote.

The Seven Seas

"The Seven Seas" is a colloquialism that roughly means "all of the world's oceans." You would figure that whomever coined the phrase must have counted something, but what? There are literally dozens of "Seas" around the world. Maybe it's simply a matter of "The Eighty-four Seas" (or whatever) just not having the same poetic impact of "Seven Seas." Someone suggested the term refers to the constituent seas of the Mediterranean Sea — like the Aegean Sea, the Ionian Sea, the Adriatic Sea, and so forth. Trouble is, my atlas shows the Mediterranean contains at least eight lesser seas, so forget that.

Then there are the oceans. My atlas shows the Arctic Ocean, The North and South Pacific Oceans, The North and South Atlantic Oceans, The Indian Ocean, and . . . that's it. Six. Some people speak of the "Antarctic Ocean" or the "Southern Ocean" (same water body), which would make seven. Indeed, there is a lot of water surrounding Antarctica. Its degree of coldness makes it ecologically distinct from adjacent oceans and is the prime rationale for separate status. But neither "Antarctic Ocean" nor "Southern Ocean" has ever achieved formal status, and therefore does not appear on world maps — at least not ones that take their reputations seriously. So what about "The Seven Seas?" My advice is that you simply say "the oceans."

The Flat Earth Society

Some people just don't get it. Which is to say yes, there really is a Flat Earth Society. See for yourself. Get on the Internet and do a key word search. All sorts of things will pop up under "Flat Earth Society" including an honest-to-goodness organization by that name. Many (most?) members are motivated by literal interpretation of Scripture. In the Gospels (Matthew 4:8; Luke 4:5), for example, Satan tempts Jesus by taking Him to the top of a very high mountain to see "all the kingdoms of the earth." That would be theoretically possible on a flat Earth, but not on a sphere-like one.

There's a certain logic to the flat Earth idea. Basic personal experience argues for it. And despite everything you know and trust about gravity, there's still something odd about being upside down at the South Pole, isn't there? Bona fide flat-Earthers feel the same way. They also believe all that photography from space showing a sphere-like Earth is a hoax — ditto for the space shuttle, at least with respect to orbiting the earth.

In any academic discipline, dissent and contrary thinking have a funny way of proving healthy. Goodness knows how many incredible discoveries have been made because somebody had the wisdom and the courage to go against the flow. But The Flat Earth Society? With all due respect to the faithful, it's time to find another conspiracy theory.

"The Continent"

Europe is sometimes called The Continent. It's one thing that Europe is even referred to as a continent, but *The* Continent? Puh-leaze! Europe is nothing more than a peninsula of Asia.

"Europe" comes from Europa, who, in Greek mythology, was a Phoenician princess who got carried off by Zeus in the form of a white bull, and by him became the mother of Minos, Rhadamanthus, and Sarpedon. No wonder people are mixed up about this "Europe" business. Actually, it was the ancient Greeks who coined the name and first placed it on their maps. And indeed in those days travel between, say, Southern Europe and Central Asia took so long and involved such perils that, for all intents and purposes, they might as well have been separate land masses. Culturally, Europe and Asia were separate worlds, too. But physically separate? Not on my map, nor even on that of the ancient Greeks. So is Europe a continent? As a matter of standard usage, yes. As a matter of geographic reality, no. And *The* Continent? Sounds to me like a matter of self-esteem — or lack thereof. As for you, forget about it.

Chapter 22

Ten Great Geographical Web Sites

Geography-related Web sites abound in cyberspace — why do you think they call it the *World* Wide Web? The problem isn't finding ten good sites, but instead having so many from which to pick! In most cases, the sites presented here were chosen for their collective breadth rather than their individual greatness. If you have not already done so, I encourage you to check out Chapter 19. Each organization listed there maintains a Web site, some of which, had they not already been mentioned, would surely be included in this chapter.

The About.com Geography Page

Arguably, this is The Mother of All Geography Web Sites (`http://geography.about.com`). I like to equate it to a huge shopping mall full of stores devoted to different aspects of geography. You want maps? They got maps. Plus info on atlases, cartography, census data, climate, country facts, cultural geography, disasters and hazards, geo-fun, geo-games, geo-humor, geo-education, GIS, GPS, geographic terms, homework help, latitude and longitude, photos, physical geography, river and streams, streets and roads, time zones, topographic maps . . . I can go on and on! When you're done with that, you can take their weekly quiz, participate in the weekly Geography Chat, or follow any one of the variety of links.

Digitalglobe.com

DigitalGlobe (www.digitalglobe.com) is one of several companies and agencies involved in satellite monitoring of Earth's surface that offer sample images or even catalogs of images for online perusal and purchase. While several galleries are available on this Web site, the Application Tour is especially recommended for the novice geographer. Here you can access a variety of image types that pertain to agriculture, mapping, environmental issues, emergency monitoring, and forestry applications. Each image or set of images is accompanied by a paragraph-length description.

Mapquest.com

Mapquest.com (http://www.mapquest.com) may be the best source of general-purpose maps on the Internet. Its world atlas allows you to zoom in on fairly detailed country maps that are accompanied by comprehensive fact sheets. For the U.S., you can zoom in on similar state-level resources, as well as maps of cities, towns, and street addresses, each of which is complemented by an aerial photograph and topographic sheet. Separate maps are available for U.S. national parks, scenic drives, and airport terminals. For North America and Europe, you can access the site's road trip planner. Just punch in your origin and destination points, and the program gives you detailed driving instructions, plus a route map that can be viewed at varying scales.

Perry-Castañeda Library Map Collection

The Perry-Castañeda Library Map Collection, University of Texas at Austin (http://www.lib.utexas.edu/maps), is home to more than 250,000 maps of all parts of the world. A representative cross-section numbering about 5,000 are available on this site. On the home page, you can click on various world regions to access map menus that pertain to different countries. In many instances, this leads to another menu that offers a mix of contemporary and historical maps, some centuries old. Maps from various journalistic sources pertaining to current and recent events also are present. On the main menu, "Finding Information" provides links to other map sources and related materials.

UNFAO

The official Web site of the United Nations Food and Agricultural Organization (`http://www.fao.org`) is a treasure trove of statistics on global — you guessed it — food and agriculture. Here you can find country-level data on all kinds of agricultural production, plus general economics, fisheries, forestry, and nutrition. An impressive supporting cast of subjects includes contemporary information on sustainable development, desertification, ethics in food and agriculture, gender and food security, biotechnology in food and agriculture, and other important items related to economic development and human well-being. In addition to these regular menu clickables, the homepage typically contains at least one "in-depth focus" essay on a timely topic (such as how the African AIDS epidemic is threatening agricultural production and future sustainable development) and current news items related to food production in different countries.

The U.S. Department of State's Geographic Learning Site (GLS)

Devoted primarily to political geography, this Web site (`http://www.geography.state.gov`) is typical of the genre that is designed with teachers and students in mind. From the home page, you can go to sections on traveling and living abroad, facts and information about countries and regions, current international topics and issues, and history and culture around the world. Interactive and historical maps are available, as well as topical ones that complement the sections mentioned. When you tire of the serious stuff, you can try your hand at one of the games or quizzes. Also, check out the links that, in addition to literally giving you the world, lead you to a host of statistics on demographics, trade, and business at home as well as abroad.

The Virtual Geography Department

Based at the University of Colorado, this site (`http://www.colorado.edu/geography/virtdept/contents.html`) is designed to help college and university geography departments share a wide range of information including the structure of their degree programs, course syllabi, modules and materials for classroom use, and online courses. For the novice geographer who may be considering "taking it to the next level," this is a great place to peruse the content of academic geography to see whether or not it looks right for you. Think of it as an online catalog for the potential college geography student or geography major.

Worldbank.org

The official Web site of the World Bank (http://www.worldbank.org) is a great source of information about conditions and trends in developing nations. From the main menu you can navigate to a series of tabular data on individual countries or the regions of which they are part. You can also read up on a number of regional initiatives — perhaps the most intriguing of which is the Indigenous Knowledge program, which seeks to incorporate traditional local know-how into development projects. Click on Development Topics and up pops about 25 issues that range from AIDS to environment to water supply. Click on Kids and Schools to access teaching modules and classroom resources for teachers, and reference material for students.

WorldClimate.Com

WorldClimate.com (http://www.worldclimate.com) offers climate data for 85,000 locations around the world. Want information on, say, Madrid, Spain? Just type in Madrid, click on Search, then click on the Madrid that happens to be in Spain (as opposed to the one in Perkins County, Nebraska), and you got data. Average monthly rainfall (in inches and millimeters) and average monthly temperature (in Fahrenheit and Celsius) data are available for most sites. Also available is information on the source(s) of the data, the precise period of time over which the data was averaged, and a map that shows the location of the weather station where the data was collected. This is an excellent resource for teachers and students of physical geography, as well as for the prospective traveler who wants advice on what to pack.

World Resources Institute

The World Resources Institute (http://www.wri.org) seeks to educate people "to live in ways that protect Earth's environment and its capacity to provide for the needs and aspirations of current and future generations." Toward that end, WRI maintains a robust, informative, and well-organized Web site. On their homepage, click on Global Topics or Earth Trends to access information on agriculture and food, coastal and marine systems, forests and grasslands, population and human well-being, and other topics related to global resources and ecological linkages. Information formats include feature articles, data tables, country profiles, maps, and searchable databases.

Index

• •

Geography For Dummies®

Geography For Dummies®

Geography Terms to Keep You in the Know

Axis: An imaginary line around which Earth rotates.

Carrying capacity: The number of people that can be supported by available resources.

Continental shelf: Raised ocean bottom along the margins of continents.

Diffusion: The spread of culture traits.

Equator: An imaginary line that is equidistant between the two poles.

Global Positioning System (GPS): A method that uses satellite information to determine accurate location on Earth.

Gradation: The wearing down of earth's surface.

Graticule: The Earth's grid system composed of lines of latitude and longitude.

High-pressure system: A region of cool descending air, usually characterized by clear skies and low humidity.

Latitude: Lines on the Earth's grid that run east to west.

Longitude: Lines on the Earth's grid that run north to south.

Low-pressure system: A region of warm ascending air, usually characterized by clouds and storms.

Natural increase: Population change that occurs when birth rates exceed death rates.

North and South Poles: The points where the axis intersects Earth's surface.

Plate tectonics: The theory that the hard outer earth is divided into plates that move relative to each other.

Prime Meridian: The zero-degree line of longitude.

Copyright © 2002 Wiley Publishing, Inc.
All rights reserved.

Item 1622-1.

For more information about Wiley,
call 1-800-762-2974.

For Dummies: Bestselling Book Series for Beginners